The EXOTIC PET SURVIVAL MANUAL

The EXOTIC PET SURVIVAL MANUAL

David Alderton

A Quarto Book

First U.S. edition published in 1997 by Barron's Educational Series, Inc.

All inquiries should be addressed to:
Barron's Educational Series, Inc.
250 Wireless Boulevard, Hauppauge, NY 11788

Alderton, David, 1956–
The exotic pet survival manual: a comprehensive guide to keeping snakes, lizards, other reptiles, amphibians, insects, arachnids, and other invertebrates/David Alderton.
p. cm.
"A Quarto book" — T. p. verso.
Includes – index.
ISBN 0-8120-9797-1
1. Reptiles as pets. 2. Amphibians as pets. 3. Invertebrates as pets. I. Title.
SF453. R4A43 1997
639. 3—dc21 96–47001
CIP

Designed and produced by
Quarto Publishing plc, The Old Brewery,
6 Blundell Street, London N7 9BH

SENIOR EDITORS Gerrie Purcell, Michelle Pickering
EDITORS Ralph Hancock, Jean Coppendale
INDEXER Dorothy Frame
SENIOR ART EDITOR Elizabeth Healey
DESIGNER James Hitchens
ILLUSTRATORS Janos Marffy, Wayne Ford
PICTURE RESEARCH MANAGER Giulia Hetherington
ASSISTANT ART DIRECTOR Penny Cobb
ART DIRECTOR Moira Clinch
EDITORIAL DIRECTOR Pippa Rubinstein

Typeset by Central Southern Typesetters, Eastbourne
Manufactured in Singapore by Eray Scan Pte Ltd.
Printed in Singapore by Star Standard Industries (Pte) Ltd.

PUBLISHERS' NOTE

The publishers would like to emphasize that some species included in this book produce toxic secretions or can cause harm with bites etc., and are therefore potentially dangerous to humans. Information and recommendations are given without any guarantees on the part of the author, consultant, and publisher, and they cannot be held responsible for any unforeseen circumstances.

CONTENTS

Introduction

INTRODUCTION 6 — HOW TO USE THIS BOOK **7** REPTILES **8** AMPHIBIANS **10** INVERTEBRATES **12**

THE HOME VIVARIUM 14 — HEATING & LIGHTING **16** DESERT **18** SAVANNAH **20** TEMPERATE WOODLAND **22** TROPICAL WOODLAND **24** SEMI-AQUATIC **26** TROPICAL AQUATIC **28**

FEEDING 30 — HERBIVOROUS DIETS **31** INSECTIVOROUS CREATURES **32** ANIMAL FOOD **33** SUPPLEMENTS **33**

HEALTH CARE 34 — FUNGAL DISEASE **35** SALMONELLOSIS **35** LUMPS & BUMPS **36** RED LEG **36** MOUTH ROT **37** VITAMIN & MINERAL DEFICIENCY **37**

BREEDING 38 — THE BREEDING PERIOD **39**

CONSERVATION CONTROLS 40 — ENDANGERED SPECIES **40** DANGEROUS ANIMALS **42**

Directory

DIRECTORY 44 — HOW TO USE SYMBOLS **45**

REPTILES 46 — LIZARDS **48** SNAKES **68** TORTOISES & TURTLES **88**

AMPHIBIANS 100 — FROGS & TOADS **102** SALAMANDERS & NEWTS **122**

INVERTEBRATES 138 — INSECTS **140** SPIDERS **146** MILLIPEDES **150** SCORPIONS **151**

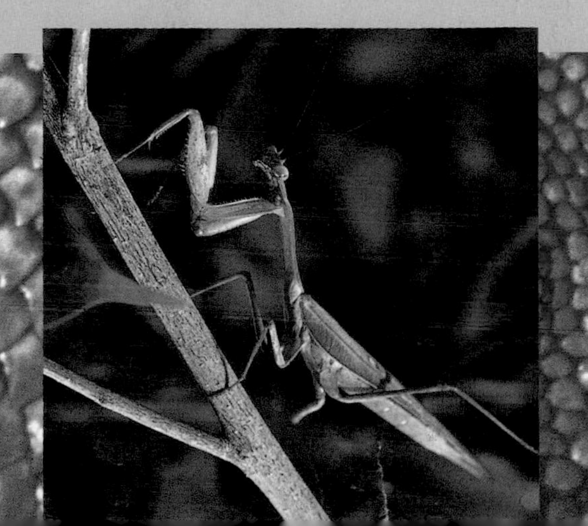

152 **TIPS**

154 **GLOSSARY**

156 **INDEX**

160 **ACKNOWLEDGMENTS**

Introduction

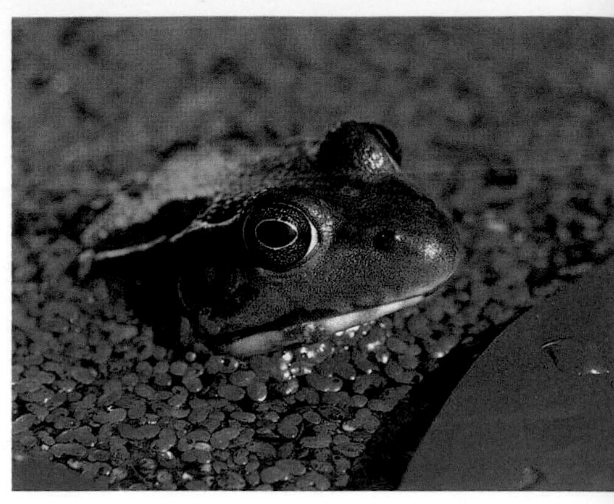

Reptiles, amphibians and invertebrates are to be found in a variety of habitats, both on land and in water.

The keeping and breeding of reptiles, amphibians and invertebrates has grown rapidly in popularity over recent years. These creatures may look exotic, but many of them can be kept quite easily, even in apartments where it would be impractical and unkind to keep a traditional pet such as a dog or a cat.

It generally takes less time to look after these creatures, too, and they are more adaptable by nature. They do not need to be fed at a set time every day, nor do they present such problems when vacations come around. Usually their enclosure or vivarium can quite easily be moved elsewhere temporarily, to a friend or neighbor.

Once established in their quarters, exotic pets normally remain healthy, and seldom fall ill unless their needs are not being correctly met. An ever-increasing range of specialist equipment is available to provide them with the best possible surroundings. Breeding of many species is also becoming commonplace. Specially formulated diets and other nutritional aids are also simplifying the care of these creatures and contributing to successful breeding in the home.

Although many exotic creatures do not enjoy close contact with those looking after them, there are a number of species that will become sufficiently tame to feed from your hand – for example, several kinds of amphibians. Some lizards and snakes can also be handled regularly, once they are used to being picked up.

There are a number of questions you should ask yourself before deciding upon a specific exotic pet, to be sure of making the right choice from the outset. If you dislike the thought of feeding them with dead rodents, for example, then choose a vegetarian or insectivorous species.

The space you have available is an important factor, because some exotic pets, particularly lizards such as the green iguana, quickly grow to quite a large size, and are soon likely to require much bigger accommodation. Do you want a tame companion which can be handled out of its quarters, or would you prefer a tropical vivarium where you can simply enjoy the plants and animals, rather as you might with an aquarium? What sort of budget do you have available? These are all questions that will shape your choice in the first instance.

This book covers a wide selection of the species that are likely to be available. The first part of the book explains how to set up a suitable environment for these creatures, based on their native habitats, and explains their general requirements, including health care and breeding. More specific information on breeding is given under the individual entry on each creature. Trouble-shooting tips and a glossary of relevant terms complete the book.

HOW TO USE THIS BOOK

The opening pages of this book explain topics such as housing and general care. The directory section then gives detailed individual coverage of more than 140 species, grouped initially into broad general categories such as reptiles, amphibians and invertebrates. These categories are then sub-divided into smaller groups in each case. For instance, lizards, snakes, tortoises and terrapins are to be found in the section on reptiles. Frogs and toads, plus newts and salamanders feature under the amphibian heading, while invertebrates range from stick insects and praying mantids to scorpions and spiders.

Introduction
Current, practical advice on looking after the featured creatures can be found on these pages.

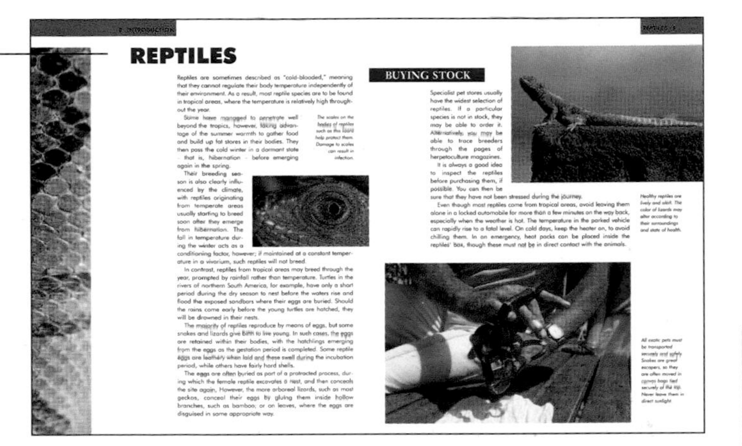
REPTILES

BUYING STOCK

Glossary
Clear, concise explanation of key words arranged in alphabetical order.

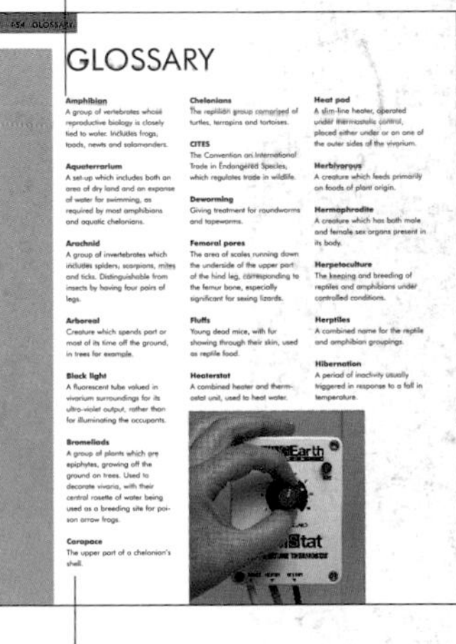
GLOSSARY

Gives useful coverage of anatomical and physiological definitions.

Directory
Standard layout enables you to compare the needs of the different species. A separate section details reproductive data.

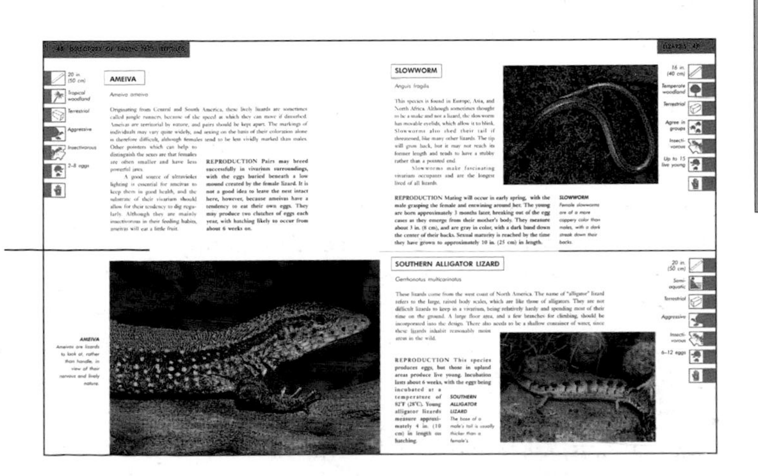
AMEIVA

SLOWWORM

SOUTHERN ALLIGATOR LIZARD

REPTILES

Reptiles are sometimes described as "cold-blooded," meaning that they cannot regulate their body temperature independently of their environment. As a result, most reptile species are to be found in tropical areas, where the temperature is relatively high throughout the year.

Some have managed to penetrate well beyond the tropics, however, taking advantage of the summer warmth to gather food and build up fat stores in their bodies. They then pass the cold winter in a dormant state – that is, hibernation – before emerging again in the spring.

The scales on the bodies of reptiles such as this lizard help protect them. Damage to scales can result in infection.

Their breeding season is also clearly influenced by the climate, with reptiles originating from temperate areas usually starting to breed soon after they emerge from hibernation. The fall in temperature during the winter acts as a conditioning factor, however; if maintained at a constant temperature in a vivarium, such reptiles will not breed.

In contrast, reptiles from tropical areas may breed through the year, prompted by rainfall rather than temperature. Turtles in the rivers of northern South America, for example, have only a short period during the dry season to nest before the waters rise and flood the exposed sandbars where their eggs are buried. Should the rains come early before the young turtles are hatched, they will be drowned in their nests.

The majority of reptiles reproduce by means of eggs, but some snakes and lizards give birth to live young. In such cases, the eggs are retained within their bodies, with the hatchlings emerging from the eggs as the gestation period is completed. Some reptile eggs are leathery when laid and these swell during the incubation period, while others have fairly hard shells.

The eggs are often buried as part of a protracted process, during which the female reptile excavates a nest, and then conceals the site again. However, the more arboreal lizards, such as most geckos, conceal their eggs by gluing them inside hollow branches, such as bamboo; or on leaves, where the eggs are disguised in some appropriate way.

BUYING STOCK

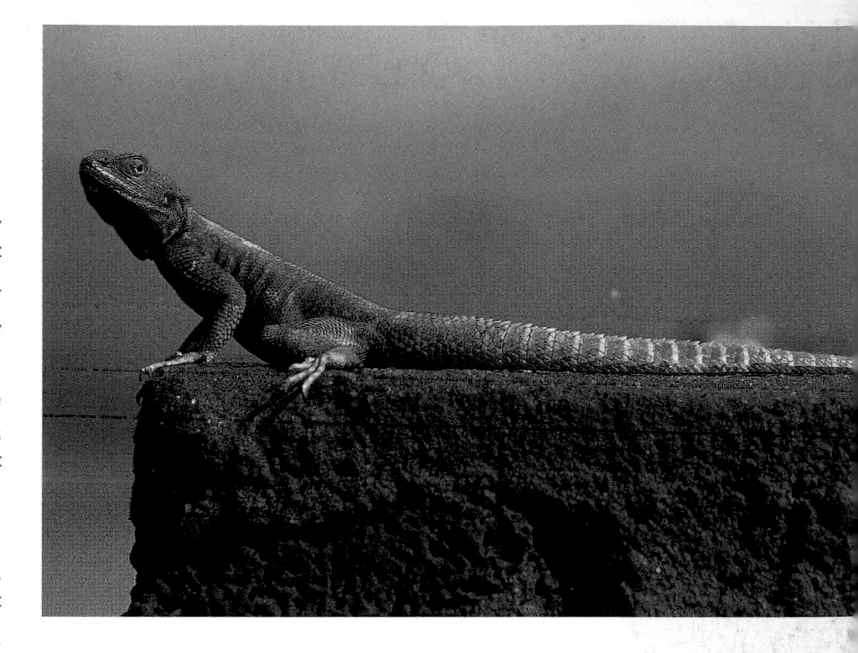

Specialist pet stores usually have the widest selection of reptiles. If a particular species is not in stock, they may be able to order it. Alternatively, you may be able to trace breeders through the pages of herpetoculture magazines.

It is always a good idea to inspect the reptiles before purchasing them, if possible. You can then be sure that they have not been stressed during the journey.

Even though most reptiles come from tropical areas, avoid leaving them alone in a locked automobile for more than a few minutes on the way back, especially when the weather is hot. The temperature in the parked vehicle can rapidly rise to a fatal level. On cold days, keep the heater on, to avoid chilling them. In an emergency, heat packs can be placed inside the reptiles' box, though these must not be in direct contact with the animals.

Healthy reptiles are lively and alert. The color of lizards may alter according to their surroundings and state of health.

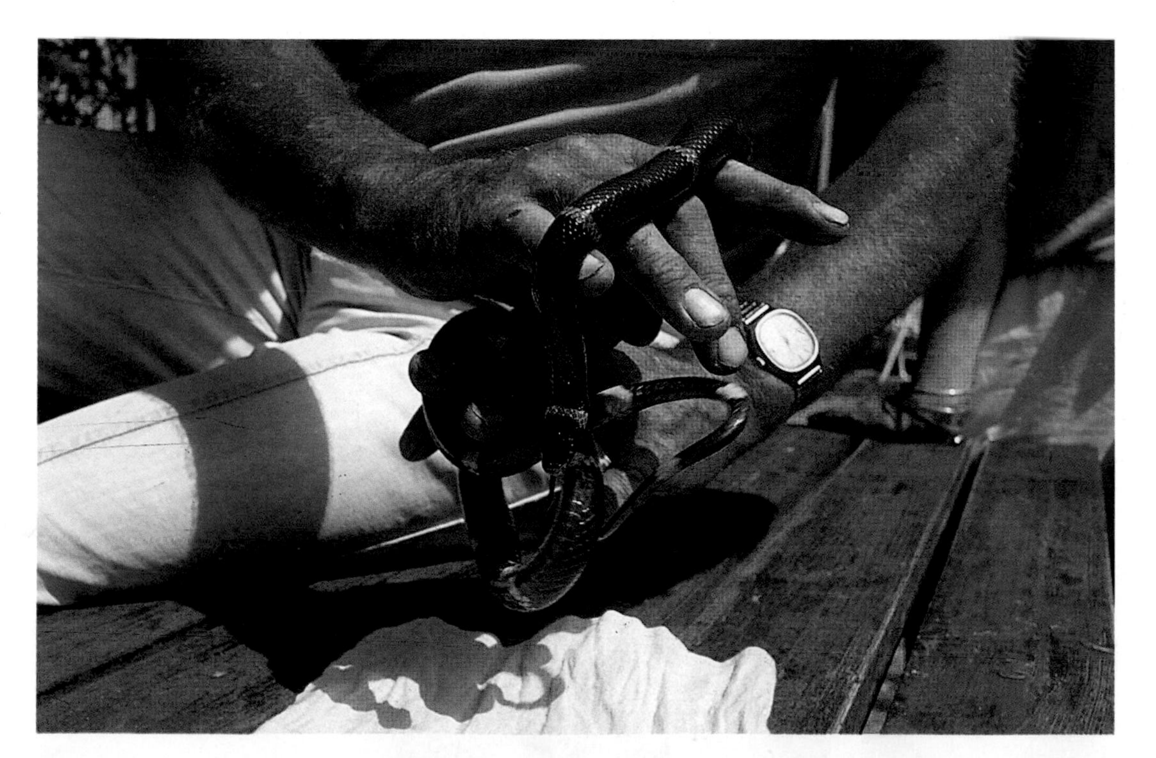

All exotic pets must be transported securely and safely. Snakes are great escapers, so they are often moved in canvas bags tied securely at the top. Never leave them in direct sunlight.

AMPHIBIANS

Amphibians are the most primitive of all quadrupeds, and are the first group of vertebrates to have evolved to live on land. They are found in damp places since they generally have to breed in water. Their name comes from the Greek words *amphi*, "on both sides," and *bios*, "life," because they live both on land and in water. Some amphibians do live and breed exclusively in water, however. A few species can even breed in their immature larval form – a process known as neoteny.

Most amphibians have a two-stage reproductive process, with the female laying her jelly-like eggs in water. These hatch into larvae, often known as tadpoles, which spend their early life in water. Gradually they start to metamorphose into adult amphibians. Their limbs begin to develop, and in the case of tail-less species such as frogs and toads the tail shortens and finally disappears, and they start to breathe air rather than relying on external gills.

One of the less well known features of amphibians is that some species look after their eggs, and even their young.

Tree frogs have enlarged pads at the tip of their toes which help them to climb vertical trunks and plant stems.

Many amphibians can swim well, and will hunt for food in water, usually feeding on various invertebrates

It has recently become clear that some amphibians care for their eggs, and even their offspring in some cases. Many female newts, for example, go to great lengths to hide each of their eggs beneath a leaf of an aquatic plant. Some frogs carry their eggs around with them, and may even provide food for their tadpoles in due course.

It can be difficult to sex amphibians visually, and so purchasing a small group is the best way of ensuring that they will breed. Some species can be obtained as tadpoles, which provides a cheaper option than starting with mature stock.

Amphibians have sensitive skins, which are easily damaged. They must be handled carefully, and never with dry hands, as this could strip the protective mucus off their bodies, leaving them open to bacterial infection which may prove fatal.

The bright colors of many amphibians appear highly attractive to our eyes, but in reality they are a warning, indicating the presence of toxic chemicals in the skin, which serve to protect them. This is another reason for handling amphibians with care. Thin disposable gloves should be worn for handling the more toxic species, such as the poison arrow frogs, since their poison can enter the body through the smallest cut or abrasion.

Amphibians are prone to desiccation, which is why they generally live in moist surroundings and rarely stray far from water.

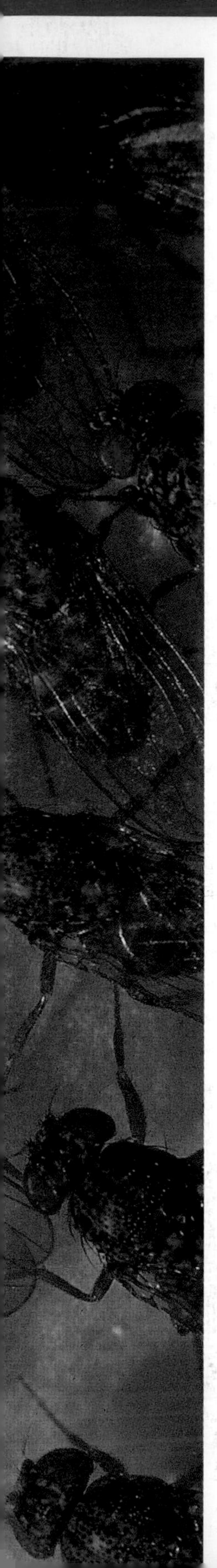

INVERTEBRATES

This general name includes insects, spiders, and scorpions, many of which are popular pets. The bizarre or colorful appearance of some species adds to their appeal.

These creatures tend to be shy, often hiding away for much of the day, and are often also quite fragile, so they are really display animals to be kept in a vivarium, rather than creatures which can be petted. Nevertheless, most invertebrates are quite easy to cater for, particularly the most popular species such as stick insects or tarantulas.

Breeding can prove equally straightforward, at least in the case of stick insects and giant land snails. In many species, there is no need to have pairs. Some stick insects are parthenogenetic, which means that adult females can lay fertile eggs which hatch into clones of themselves. The young stick insects, called nymphs, resemble miniature adults, growing through a series of molts.

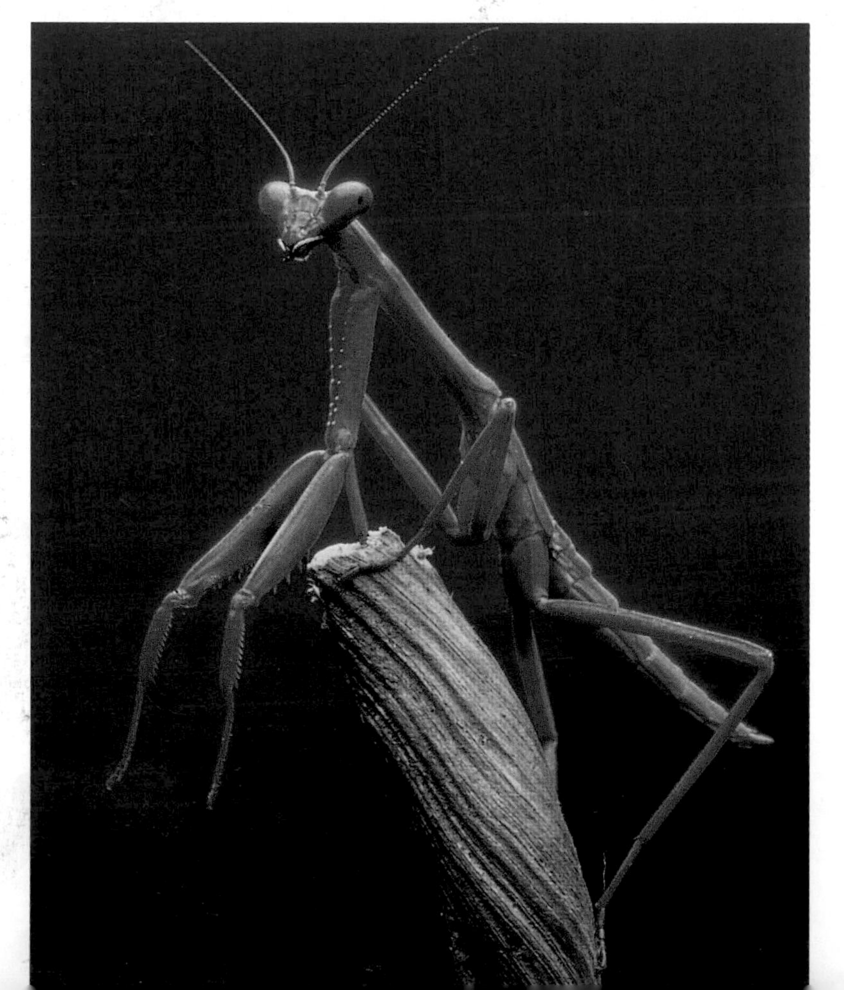

Some invertebrates, such as the praying mantis, can be highly predatory by nature, even eating members of their own species.

In some cases their body shape will alter, causing them to become more heavily disguised as they become bigger and inevitably more conspicuous. Giant land snails have a different means of reproduction, being hermaphrodite. Each snail has male and female sex organs in its body, and so two kept together will inevitably produce fertile eggs, which give rise to miniature snails.

Tarantulas, which have clearly defined sexes, practice some parental care towards their eggs, with the female keeping them in a special ootheca or egg sac. But it is the scorpions, closely related to these spiders, which show the greatest concern for their offspring, with the female actually carrying her young on her back for a time after they are born.

It is possible to purchase invertebrates at all stages in their life cycle, and seeking out spiderlings, for example, will often be considerably cheaper than purchasing adult tarantulas. You can also be certain of their age, which is especially important in the case of male tarantulas. They have a considerably shorter lifespan than females.

The reproductive habits of many invertebrates can be fascinating and the likelihood of breeding success in vivaria is quite high. Sexing is possible in many cases, as with tarantulas.

With stick insect eggs, it is important to bear in mind that a percentage are unlikely to hatch successfully, even under ideal conditions. The length of time taken for the young to emerge from their eggs can also be quite variable, so do not be in too much of a hurry to discard the eggs prematurely.

As with the other kinds of creatures covered in this book, there are hobbyist organizations which are useful not only for obtaining stock from fellow enthusiasts, but also in finding good homes for your surplus stock in due course. Such organizations often produce regular newsletters and bulletins, and may organize events such as shows as well.

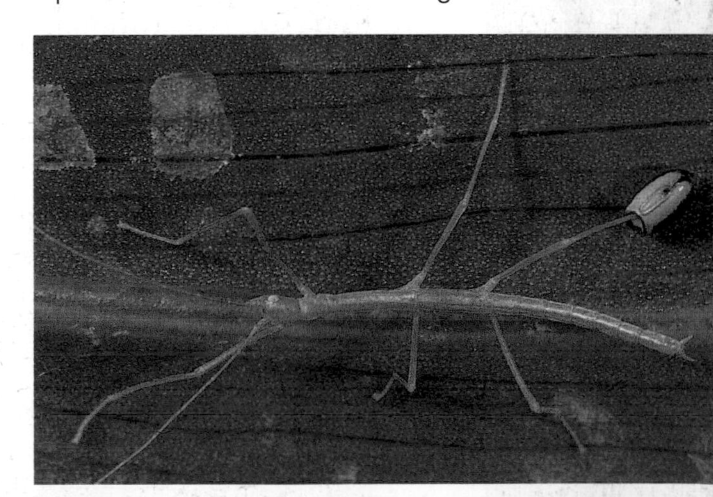

Most invertebrates have many young since, in the wild, most of these will fall prey to other creatures before reaching maturity. This hatchling stick insect still has its egg case attached to one foot.

THE HOME **VIVARIUM**

The type and size of enclosure you need depends on the species you are intending to keep. For aquatic species, a converted aquarium is recommended. This should be equipped with a special ventilated hood, which also includes fittings for lighting. Access to the interior is generally through a sliding lid in the hood. This is a relatively secure option, bearing in mind that many amphibians, lizards, and snakes can jump and climb very effectively. Small stick insects are also able to walk up vertical surfaces, such as the side of a tank, and a secure cover which fits closely over the top of the unit is essential to contain them.

Glass vivaria are fairly inexpensive, but they are much heavier than plastic containers and easily broken. It is possible to obtain acrylic vivaria for small species which do not require additional heating or lighting, complete with ventilated plastic hoods including an opening door.

Most glass vivaria today have their panels set in a flexible silicone sealant, which waterproofs them very effectively. However, if you are keeping turtles in such a container, you need to check that their sharp claws do not inflict any damage to the sealant, as the effects could be catastrophic. A layer of well washed coarse aquarium gravel will not only look more attractive than a bare base, but will also serve to protect the sealant. Cleanliness will be more difficult, however, because the gravel will need to be emptied out of the aquarium and thoroughly rinsed in a clean bucket, whenever the tank is cleaned.

The situation with turtles is further complicated by the fact that nearly all species need to come ashore at times, so there also needs to be a dry area of land under a spotlight where they can bask. It may therefore be better simply to have

Heat source protected by mesh
Ventilation grill
Ultraviolet fluorescent strip light
Hinged lid
Vertical branches set in buried concrete bases for stability
Plants of suitable size set in pots in bark substrate
Clear plastic or glass front
Digital thermometer
Moss covering over the floor
Handle on tray to facilitate cleaning

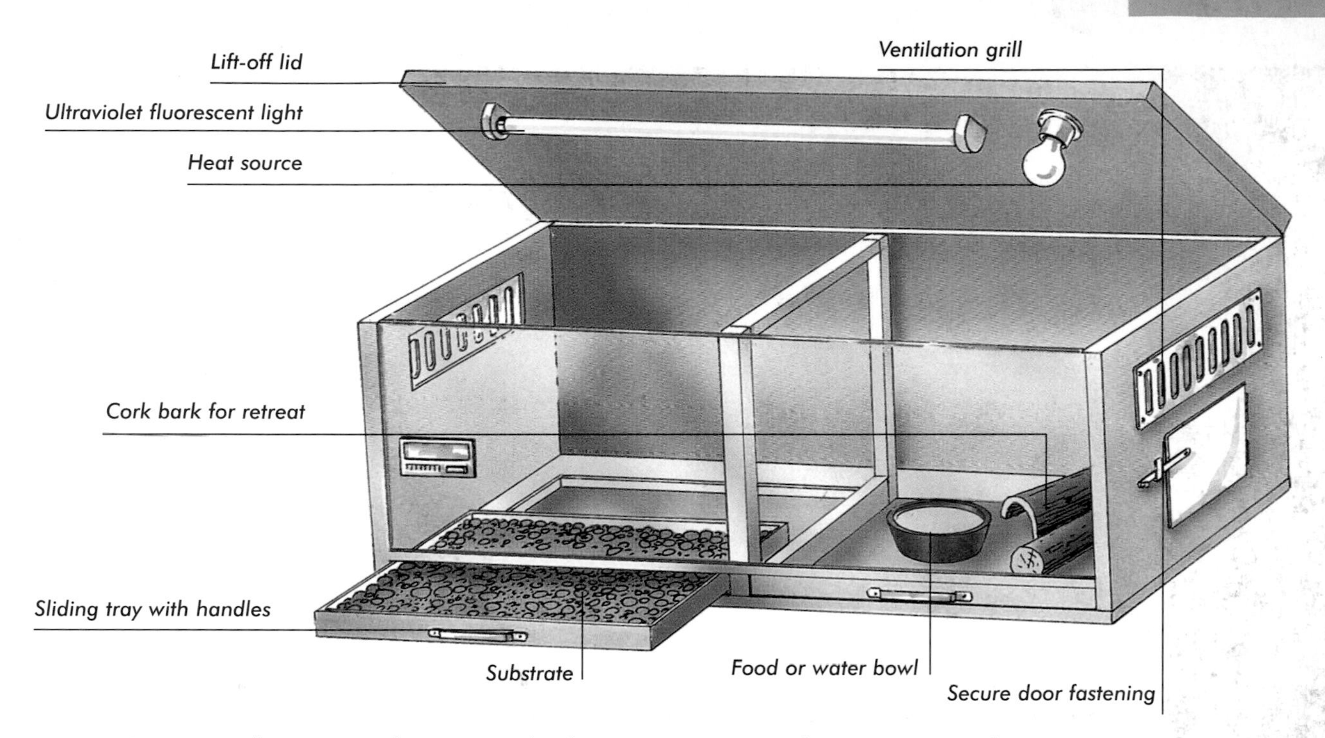

an aquatic area without any substrate, and rely on an aquatic siphon to remove the water into a bucket, then wipe the interior with paper towelling and finally refill the aquatic part of the tank. Turtles need more regular cleaning than most other reptiles.

Snakes and lizards can be kept in glass vivaria, but these units tend to lose heat quite readily. Melamine vivaria, available in a wide range of sizes, are ideal for these reptiles. It is easy to install additional heating and lighting in such a container, and the surface can be wiped over and disinfected easily.

A melamine box also gives the creatures a greater sense of security than a transparent glass unit, which is especially important for nervous species. Access is normally by sliding doors at the front of the vivarium, which will need to be kept securely closed with a lock. Both lizards and snakes can prove to be excellent escape artists, using their snouts to push back any insecure part of the vivarium and then disappearing into the room.

The location of the vivarium in the room is important. Although many reptiles in particular originate from hot areas of the world, the vivarium should never be located in front of a window. The glass will magnify the effects of the sun's rays quite dramatically, rapidly raising the temperature to a fatal level. A corner of a room, out of direct sunlight, is to be preferred. This will also lessen the likelihood of unsightly algal growth in aquatic vivaria. It is also important to set the vivarium on a secure base, particularly if it contains water, because even a partially full aquarium will weigh a considerable amount.

Follow the recommendations for the unit concerned; it may be advisable to stand it on a piece of polystyrene sheeting. This will take up any irregularities in the surface, which might otherwise exert pressure on a part of the aquarium, possibly causing it to break or leak.

It obviously helps if the vivarium is positioned in such a way that you can enjoy watching its occupants comfortably from within the room. The only other definite requirement is easy access to a double receptacle (power point) for heating and lighting. Avoid trailing wires or a mass of adaptors, which can be dangerous. The type and size of enclosure you need depends on the species you are intending to keep.

HEATING

Heating options for the vivarium – a ceramic heater with reflector surround, a spotlight, and a heat pad – are shown here.

The simplest way to heat the vivarium is to hang an ordinary incandescent light bulb over the occupants. This can give out considerable heat, but obviously in order to be effective it needs to be left on all the time, which is not desirable because the creatures will have to be maintained in permanent light, if they are to be kept warm.

A better option, which also mimics the natural increase in daytime temperature in the wild, is to have a background heat source and a small spotlight within the vivarium to create a warm place for basking. At night the lamp can be turned off, allowing the enclosure to cool down to some extent while the background heat output is maintained. The spotlight will be more effective if it has a reflector – but avoid overheating the animals.

The aim in designing the heating system should always be to provide a thermal gradient across the enclosure. This will allow the reptile to bask under the spotlight to raise its body temperature and level of activity, after which it can move away in search of food, for example, returning to bask again after it has eaten.

A heat pad connected to a thermostat can be used to provide background heating. The main function of this thermostat will be to ensure that the temperature does not fall too low. Suitable thin, flexible heat pads are available in a variety of sizes to fit either beneath or at the back of the vivarium unit, transmitting warmth through the glass, or they can be placed inside.

Mercury thermometers can be broken by large reptiles.

Great care should be taken to shield any heat source used for creatures such as geckos, which can climb and so could come into direct contact with it and be burnt, perhaps fatally. In situations where bright lighting is not required, an infra-red ceramic heater can be used to provide the localized heat source, although even with this, screening with mesh to prevent any injury to the vivarium occupants is essential.

The water in an aquaterrarium can be heated with a "heaterstat," as sold for aquaria, although it is vital to ensure this remains submerged at all times. As a further safety precaution, never place your hand in the water unless the heater is switched off. Otherwise, if there is any electrical fault, you could be electrocuted.

The water in an aquaterrarium is shallower than in a fish tank, so a compact design of heaterstat should be chosen. Wrapping the heater unit in plastic mesh will prevent amphibians in particular from inadvertently burning themselves.

Liquid crystal thermometers attached to the outside of the vivarium will provide an easy means of checking the temperature within, either on land or in the water. Having two – one at either end of the vivarium – will indicate the thermal gradient as well.

The desired temperature can be maintained in the vivarium by a thermostat. This prevents extremes of heat or cold which could be fatal.

LIGHTING

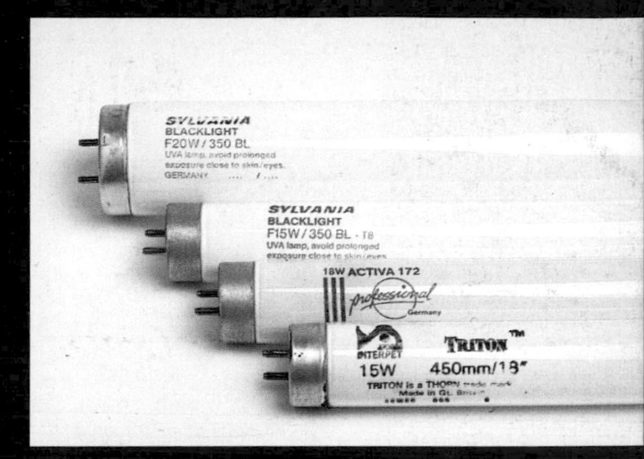

Different kinds of lighting are now available for the vivarium. Choosing the correct type can be important for the health of the occupant.

Aside from providing warmth, lighting is also essential to the well-being of many reptiles, particularly lizards and chelonians. Outdoors, the ultraviolet radiation present in sunlight falls on their skin and stimulates the synthesis of vitamin D_3 here. This vital vitamin helps the creatures to absorb vital calcium from their diet, and to use it effectively.

Reptiles that are deficient in vitamin D_3 typically have weak limbs, combined with abnormally soft shells in the case of chelonians. Young, fast-growing species such as green iguanas (*Iguana iguana*) are particularly at risk because their calcium requirement is significantly higher than that of adults.

It is possible to compensate to some degree for lack of natural vitamin D_3 with a dietary tonic, but undoubtedly the creatures will benefit greatly from being able to bask under a light that contains the appropriate wavelengths. The form known as UVA stimulates appetite and has a general tonic effect, while UVB is vital for the synthesis of vitamin D_3.

Spotlights will provide local areas of heat in the vivarium, and encourage basking in reptiles such as lizards.

Fluorescent tubes for this purpose are widely available from specialist reptile stores. They are manufactured in various lengths, to fit into hoods of the appropriate size above the vivarium. Those featuring a so-called "power twist" have a higher output of light for a given wattage than normal tubes.

It is important to remember that the strength of the beneficial wavelengths falls off sharply with increasing distance from the tube. This means that the vivarium occupants must be able to bask directly under the light, in order to obtain maximum benefit from it. A raised area here may therefore be helpful, provided that the light remains out of their reach.

Even if you allow your reptiles out of their quarters and allow them to sit on a windowsill on a warm day, they will not benefit from the sun's ultraviolet rays, because these are filtered out by glass. In areas where the climate is favorable, however, it may be possible to transfer them to an outdoor enclosure for part of the year, where they will be able to bask in unfiltered sunlight.

Even in temperate areas, it is possible to transfer terrapins to an outdoor enclosure on summer days to bask, and then bring them indoors at night so they do not become chilled. Those species of North American origin are especially susceptible to a vitamin D_3 deficiency if deprived of sunlight. The problem only shows up gradually, because the vitamin is stored in the liver.

A more concentrated source of suitable ultraviolet light is a so-called "black light" tube, which emits little visible light as the name suggests. This therefore has to be used with an ordinary fluorescent light, in order to illuminate the vivarium.

With all kinds of ultraviolet light, output declines slowly over a period of time, although the light appears to be functioning normally. The units should therefore be changed in accordance with the manufacturer's instructions, to ensure that the reptiles receive maximum benefit from them. This normally needs to be done once a year.

Heat sources of this type are fixed to the roof of the vivarium, where they must be shielded to prevent arboreal species from burning themselves.

THE TYPES OF **VIVARIUM**

DESERT

Keeping a check on the temperature is straightforward with a digital thermometer fixed to the front of the unit.

This type of environment should obviously be hot and dry. Sand would seem to be the natural substrate for a vivarium of this type, but gravel is often a better option, because sand may stick to food, and can cause digestive upsets.

It is important to provide a deep layer of substrate, because desert-dwelling reptiles are burrowing creatures by nature, spending much of their time hidden underground, certainly when the sun is at its hottest. They tend to emerge at dusk, while the air is still warm, and then return underground before the chill of the night sets in and deprives them of their ability to move quickly.

In many ways these creatures are quite hardy and adaptable, but damp conditions will not suit them. Water should, however, be provided in a shallow dish. Spotlights creating a thermometer reading of up to 104°F (40°C), are vital, but the temperature should be allowed to fall as low as 58°F (14°C) at night. Strong ultraviolet lighting is also necessary, mimicking the effect of the hot desert sun.

For burrowing purposes, hollow tubes of the appropriate size can be set at an angle in the substrate, since the substrate is unlikely to be solid enough for the reptiles to create burrows that will not collapse. For egg laying, provide a container that is partly buried in the surface of the substrate and contains moist sand. Although it is possible to decorate the vivarium in some cases with desert plants, avoid cacti with sharp spines. While these may not necessarily harm the occupants, they could injure you when you are attending to them.

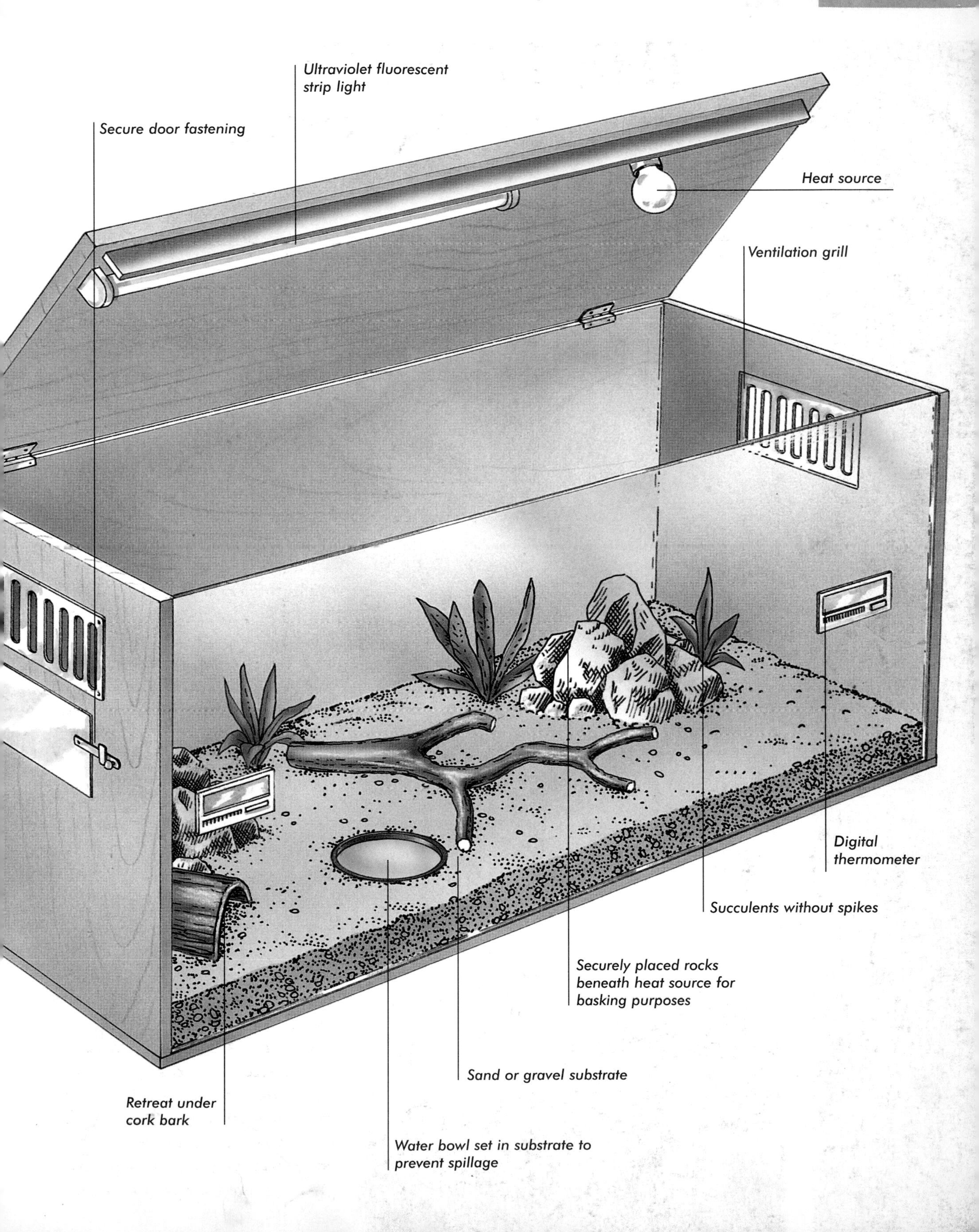
Ultraviolet fluorescent strip light
Secure door fastening
Heat source
Ventilation grill
Digital thermometer
Succulents without spikes
Securely placed rocks beneath heat source for basking purposes
Sand or gravel substrate
Retreat under cork bark
Water bowl set in substrate to prevent spillage

SAVANNAH

Many reptiles live in this type of environment, where the daytime temperature is high but less so than in desert areas. There can be long dry spells in this type of terrain, followed by torrential yet quite brief periods of rain, which often act as a stimulant to breeding, particularly for amphibians. A temperature range of 75–86 °F (24–30 °C) should be adequate in most cases.

The arrangement and choice of plants needs to be based on the species kept in the vivarium. Lizards may flatten some plants, while climbing up others. Cork bark and plants can provide retreats for many smaller reptiles and amphibians which are shy by nature. Light wood chips can be used as a floor covering.

There is more cover in a savannah setting than in desert, and retreats such as areas of cork bark and rockwork are advisable. Be sure that any such decor is securely positioned, however, to avoid injuries to the occupants caused by a falling rock. Coarse gravel will make a good substrate, although light-colored wood chips are another possibility.

You can purchase artificial decor for the vivarium, including plastic plants that will make an attractive backdrop but will not be eaten by herbivorous species. Plastic plants are also more hygienic, because they can be removed and scrubbed when necessary.

If you choose live plants, however, bear in mind that they will grow, and they should remain in proportion to the vivarium as a whole, if the overall effect is not to be spoilt. Set the plants in pots, buried in the substrate or disguised with other decor in the vivarium, as this will tend to curtail their growth. Water them as necessary about once a week.

In any case, spray the vivarium two or three times a week. Increase the humidity level by more regular spraying, and perhaps add a water flow as well, to mimic the onset of the rainy period. This should stimulate breeding if the vivarium occupants are in good condition.

CORK BARK

ROCKS

LIGHT WOOD SHAVINGS

PLANTS

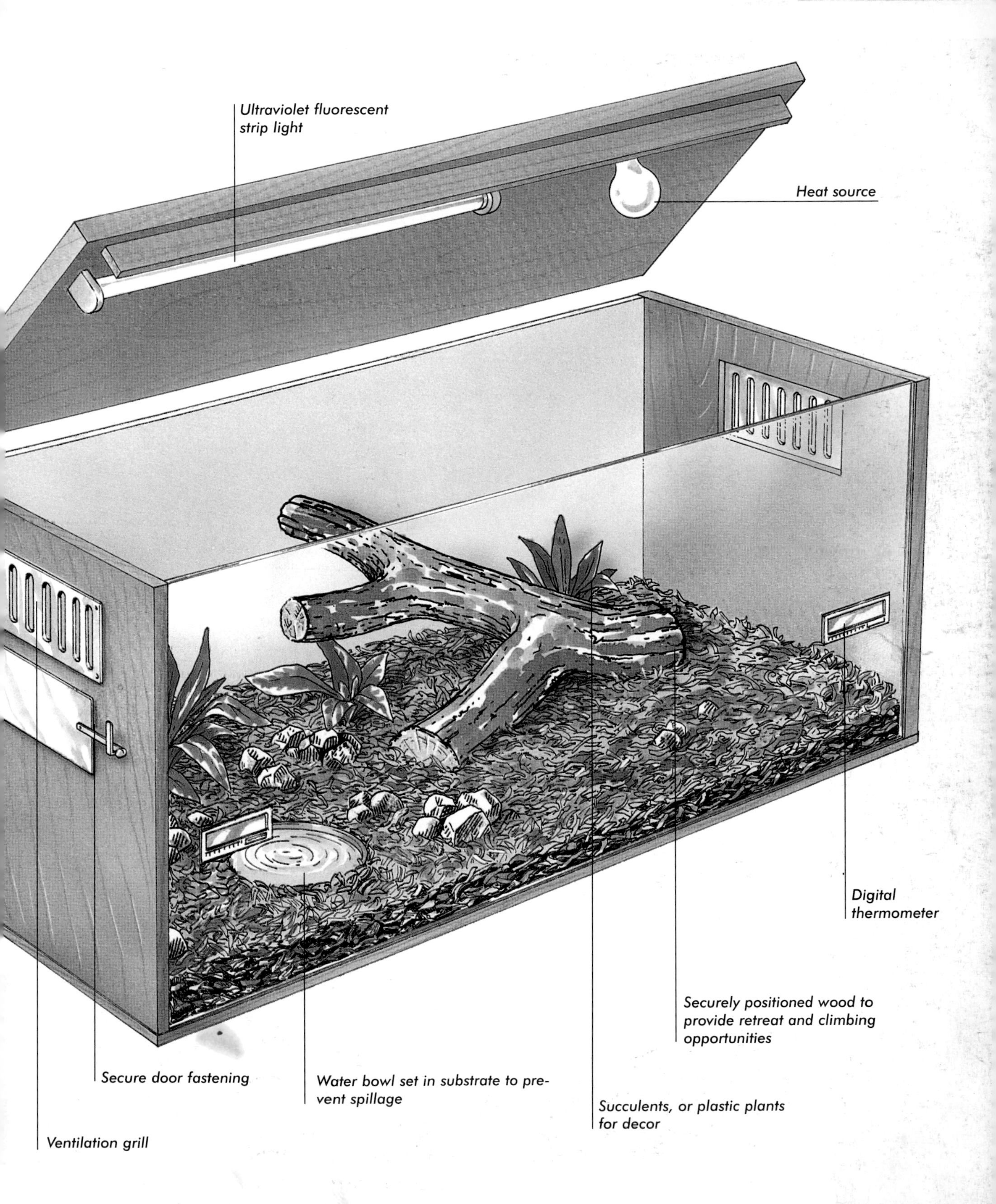

Ultraviolet fluorescent strip light
Heat source
Digital thermometer
Securely positioned wood to provide retreat and climbing opportunities
Secure door fastening
Water bowl set in substrate to prevent spillage
Succulents, or plastic plants for decor
Ventilation grill

TEMPERATE WOODLAND

This type of habitat can be home to a wide variety of species, and so the vivarium should be designed to accommodate those such as salamanders which may require relatively humid conditions, or others like some toads which prefer drier woodland surroundings. Subdued lighting is also recommended, with no searing heat lamp.

A daytime temperature of 68–75 °F (20–24 °C) will be generally suitable under these circumstances, and this should be allowed to fall as low as 50 °F (10 °C) at night. The temperature may need to be kept up by means of a low-wattage heat pad at this time if there is no overnight heating in the room. It is not a good idea to position the vivarium near or next to a radiator, as this will lower the relative humidity quite dramatically, which is likely to be counterproductive.

A slightly moist environment is the aim in this type of vivarium, but the temperature should not be much above that of the average heated room. A darker substrate is preferable in a temperate woodland setup, mimicking the natural background of the creatures which inhabit such an environment.

A layer of gravel, with bark or wood chips such as beech on top, will create a natural setting. Where there is a need to include a relatively large dish of water, then this can be set directly into the gravel. In this way it will not become overgrown with mold.

Spaghnum moss, as sold by specialist reptile stores and florists, can be included as well, preferably close to the water. It should be sprayed lightly but regularly to keep it from drying out. The moss will then provide a damp area within the vivarium, which is likely to be used by amphibians in particular. Large pieces of wood can also be incorporated, to provide surface retreats.

The temperature should be allowed to fall off in winter, as happens naturally in such surroundings. This is an important reproductive trigger for many species, which will then breed during the spring and early summer.

WOOD CHIPS

BOWL

GRAVEL

BARK CHIPS

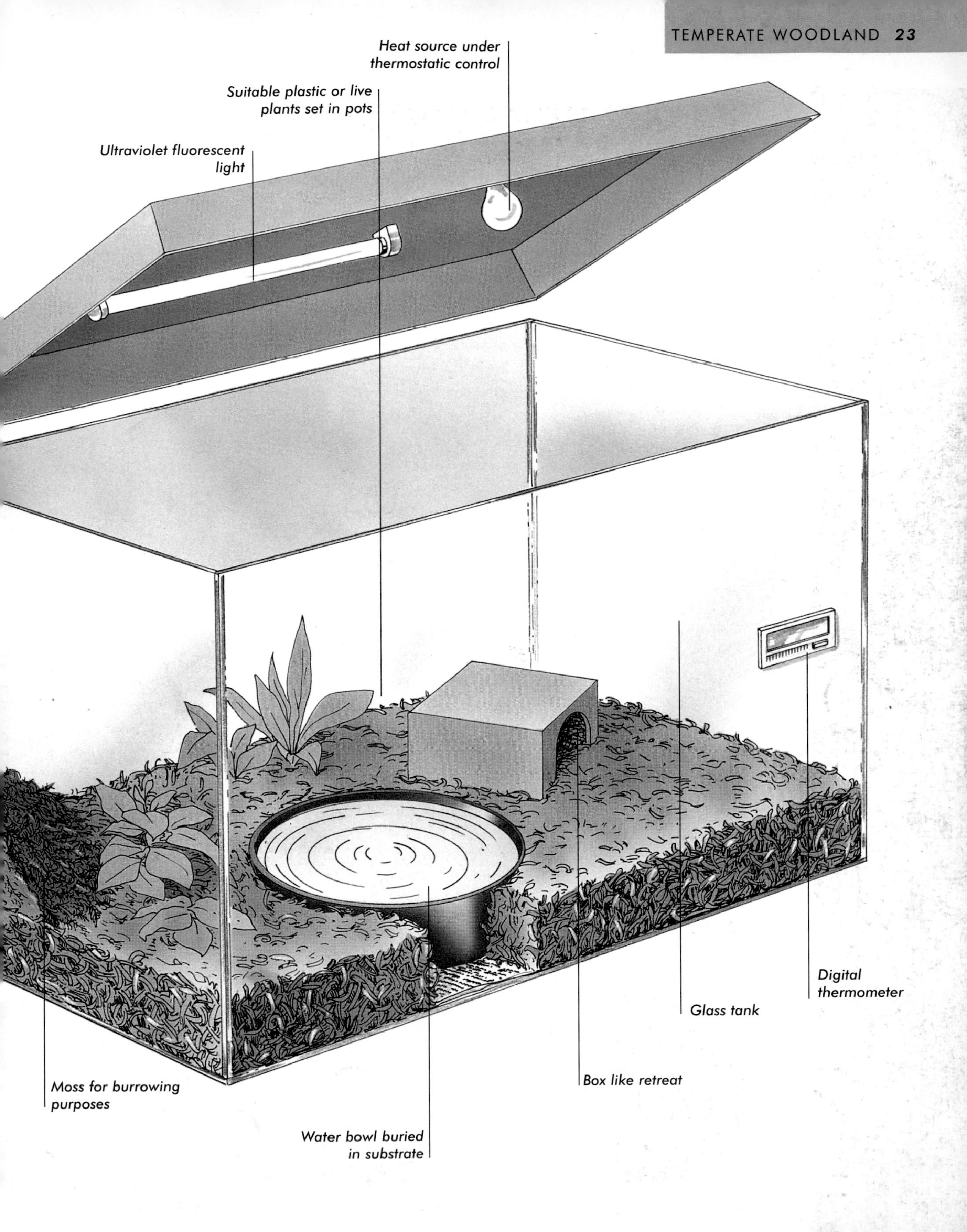
Heat source under thermostatic control
Suitable plastic or live plants set in pots
Ultraviolet fluorescent light
Digital thermometer
Glass tank
Box like retreat
Moss for burrowing purposes
Water bowl buried in substrate

TROPICAL WOODLAND

Although it is important to keep the vivarium moist, be sure to allow adequate ventilation as well; otherwise plants will dampen off and molds may develop on the sides of the vivarium. Moss can be attached to branches as can bromeliads, or be used on the floor along with bark chips and logs. These help to maintain the humidity.

Many of the creatures found in this type of habitat, including lizards, snakes, tree frogs, and tarantulas, are arboreal in their habits, and so the vivarium should be relatively tall. This also offers greater possibilities for decor. Plants can be fixed to wooden branches, or trained up them.

Aside from providing cover, these can be helpful for breeding purposes as well. Bromeliads, for example, with a cup-like rosette at the center of the plant, are used as a depository for the tadpoles of poison arrow frogs (see page 108), and can be included for this purpose in a vivarium. As epiphytes, bromeliads may be wired or tied carefully into place on to a branch of dead wood, providing color here as well.

Care needs to be taken, as some tropical woodland reptiles in particular are likely to destroy plants growing in their enclosure. Suitable plastic substitutes will be needed here.

The temperature should be maintained at a relatively high level in the tropical vivarium, ranging from about 90 °F (32 °C) during the day down to about 75 °F (24 °C) at night. This should be kept up throughout the year, in contrast to the situation in the temperate setting, since there are no distinct seasons in the tropics.

Regular spraying inside the vivarium to maintain the humidity is important to the wellbeing of the occupants, but good ventilation is equally significant. Otherwise, the interior will soon start to turn moldy. The signs will be particularly evident if there are live plants here. They will start to "dampen off," with their leaves being affected by mold, as will any buds.

BARK CHIPS

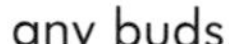

BROMELIAD PLANTS

LOGS

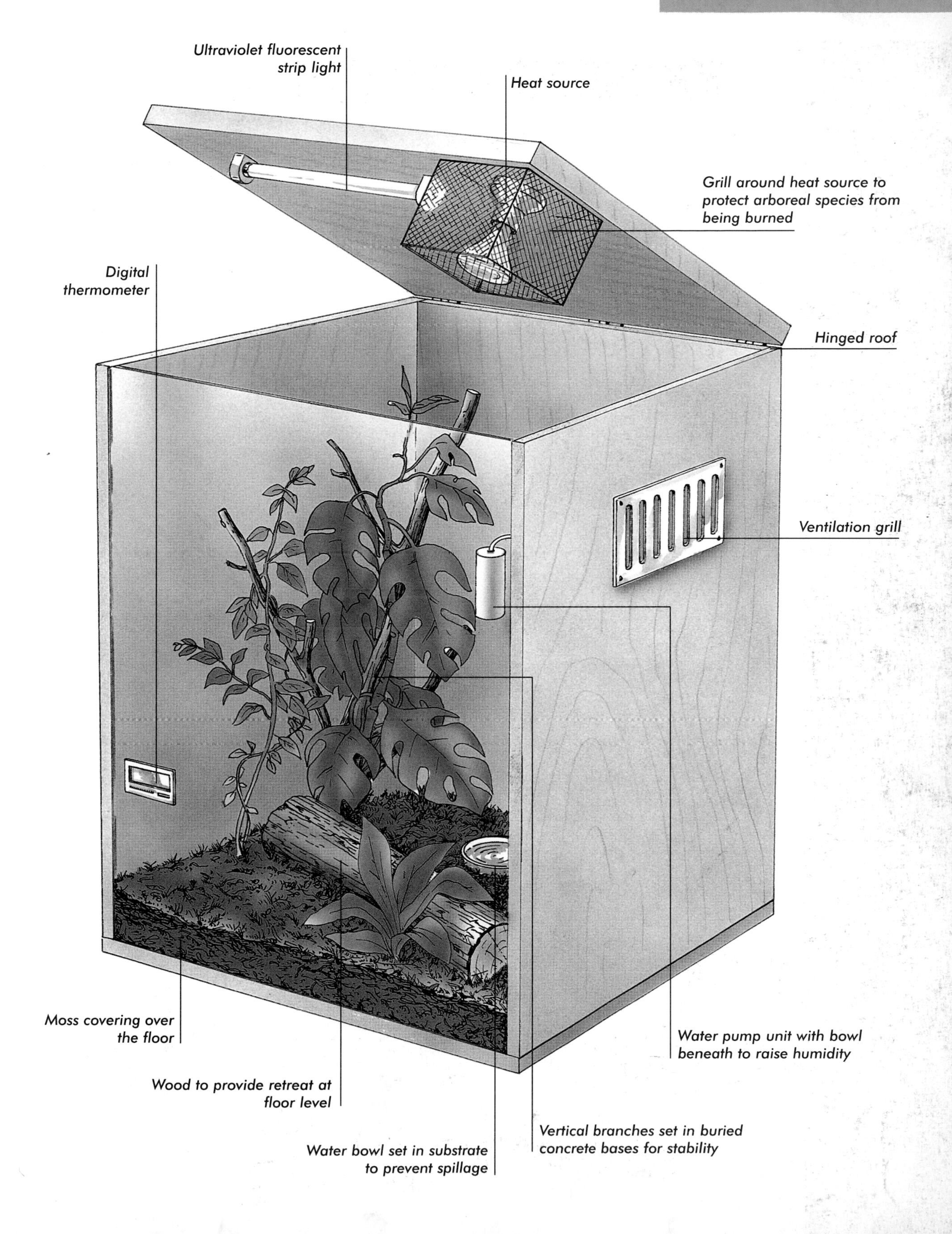
Ultraviolet fluorescent strip light
Heat source
Grill around heat source to protect arboreal species from being burned
Digital thermometer
Hinged roof
Ventilation grill
Moss covering over the floor
Water pump unit with bowl beneath to raise humidity
Wood to provide retreat at floor level
Vertical branches set in buried concrete bases for stability
Water bowl set in substrate to prevent spillage

SEMI-AQUATIC

This setup is required by most turtles, and a number of amphibians. Their needs may change through the year, however, with amphibians from temperate areas tending to become more terrestrial during the winter months, before returning to the water to breed in the following spring.

In a suitably large vivarium, which includes a sandy area with spotlights above to maintain the temperature when they are out of the water, turtles can be bred successfully. The females emerge onto land to lay their eggs. They will also bask regularly under the warmth provided by the lights. You should add damp moss for amphibians for burrowing purposes.

Land areas are important in a semi-aquatic setup, because if creatures cannot leave the water, they may develop fungal infections. The heating mat needs to be placed under the tank or attached to the side. Wash the gravel thoroughly in a colander to ensure that it is clean.

One of the most important points in a semi-aquatic vivarium is to partition the tank in such a way that the creatures can move easily from water to land and back. A suitable slope should be provided for this purpose. Avoid including rocks piles in the aquatic area, because dirt can accumulate behind them and if dislodged a rock could injure the occupants of the vivarium or break the glass.

The temperature of the water will depend on the origins of the species concerned. In the case of turtles, this should be in the region of 75–81 °F (24–27 °C), but amphibians coming from temperate areas will benefit from lower water temperatures, often with no heating required in indoor surroundings.

It is often unnecessary to include plants. Turtles will either uproot or eat plants, and their presence will also make it difficult to clean the aquatic part of the setup. When breeding amphibians, however, the addition of aquatic vegetation can encourage spawning. It also makes it easier to remove eggs, which are likely to be wrapped up with the plants, so that they can be hatched and the tadpoles reared successfully elsewhere.

GRAVEL

THERMOMETER

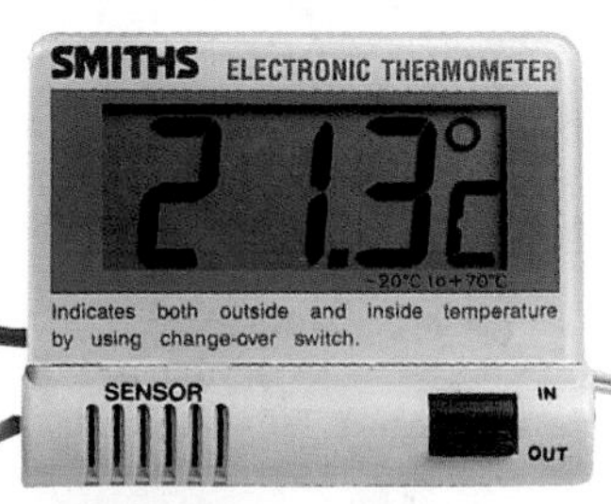

HEAT MAT

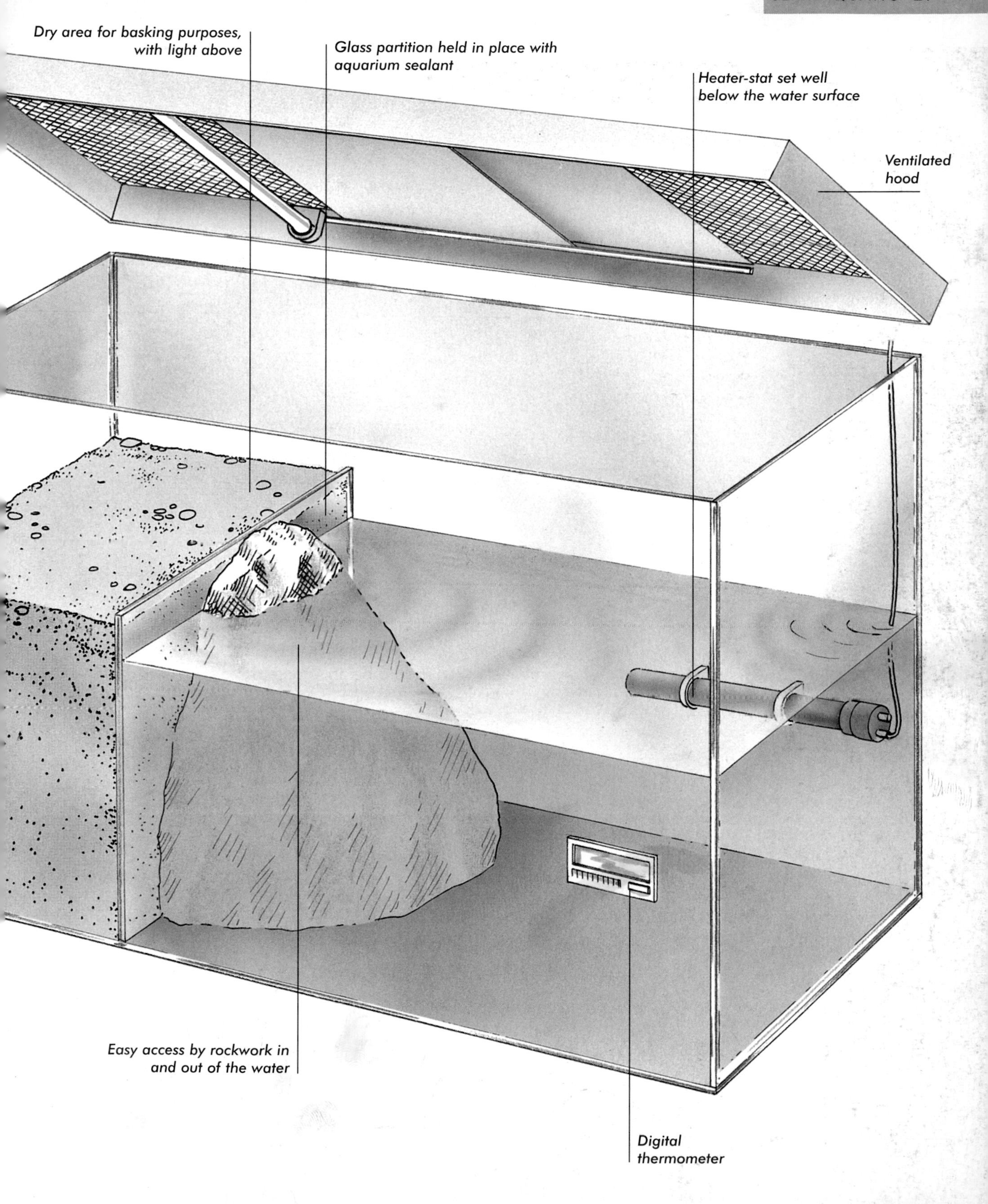
Dry area for basking purposes, with light above
Glass partition held in place with aquarium sealant
Heater-stat set well below the water surface
Ventilated hood
Easy access by rockwork in and out of the water
Digital thermometer

TROPICAL AQUATIC

Relatively few reptiles or amphibians need to be housed in an exclusively aquatic environment once they are mature. One amphibian that requires such surroundings is the axolotl, which relies on gills to extract oxygen from the water, and cannot breathe atmospheric air. While this species will not need additional heat at room temperature, other aquatic amphibians such as the African clawed frog require warm water, heated to 81°F (27°C) in this case.

An undergravel filter will keep the water clean, as in a fish tank. Other equipment produced for fish keeping, such as a heaterstat, will also be useful, although the aquarium itself should not be filled completely to the top. Lighting in the aquarium hood will illuminate the tank occupants beneath, while floating plants on the surface of the water should ensure that the light is suitably diffuse.

It is generally not a good idea to include plants in the gravel substrate at the base of the aquarium. They are likely to be uprooted by the often rather clumsy swimming habits of the reptiles and amphibians that live in this type of environment. Cleanliness is of paramount importance, and the water must also be well oxygenated, so that the provision of an air pump in the system is recommended. A dechlorinator or water conditioner, as sold for use with fish, can prove beneficial as well.

The equipment for this type of vivarium is widely used by fish keepers, although in this case the tank should not be filled right to the top. Floating plants provide valuable cover at the surface and will thrive under suitable lighting. Choose a quiet air pump – some models can be noisy.

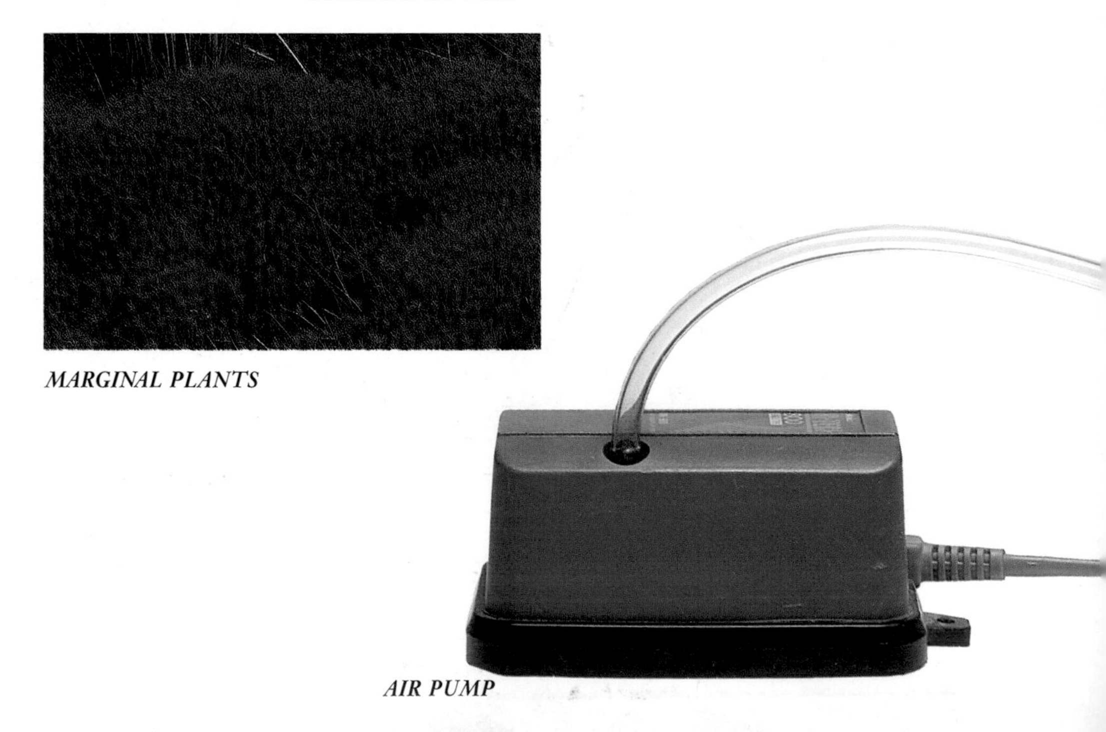

MARGINAL PLANTS

AIR PUMP

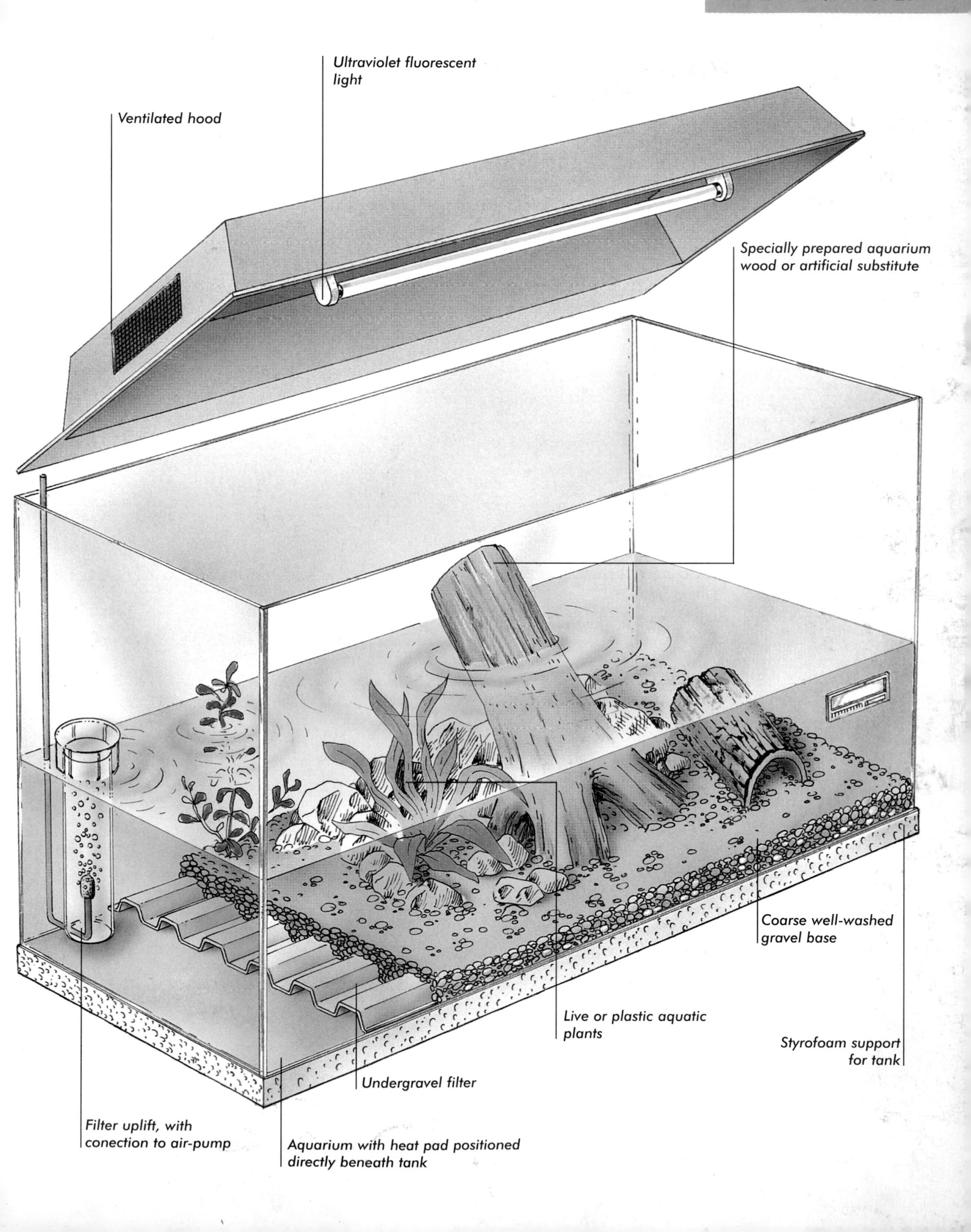
Ultraviolet fluorescent light
Ventilated hood
Specially prepared aquarium wood or artificial substitute
Coarse well-washed gravel base
Live or plastic aquatic plants
Styrofoam support for tank
Undergravel filter
Filter uplift, with conection to air-pump
Aquarium with heat pad positioned directly beneath tank

FEEDING

The actual basic dietary requirements of the creatures covered in this book do not differ significantly from our own, in that carbohydrates are needed for energy, proteins for growth and healing of injuries, and fats to provide essential fatty acids and to act as a concentrated source of energy. Fat may also be stored in the body to sustain the creatures through hibernation. Vitamins and minerals are needed in small amounts to carry out a diverse range of functions in the body. Calcium, for example, is needed for healthy bone structure, and iodine is vital for the metamorphosis of amphibians.

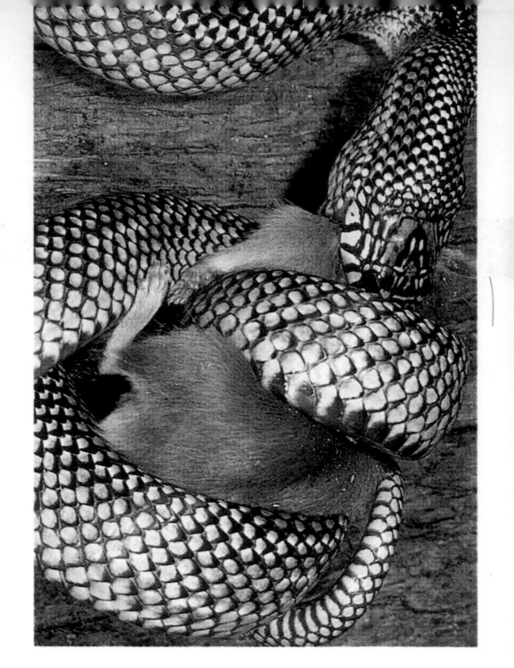

All snakes are predatory in their feeding habits. A blotched king snake is shown here.

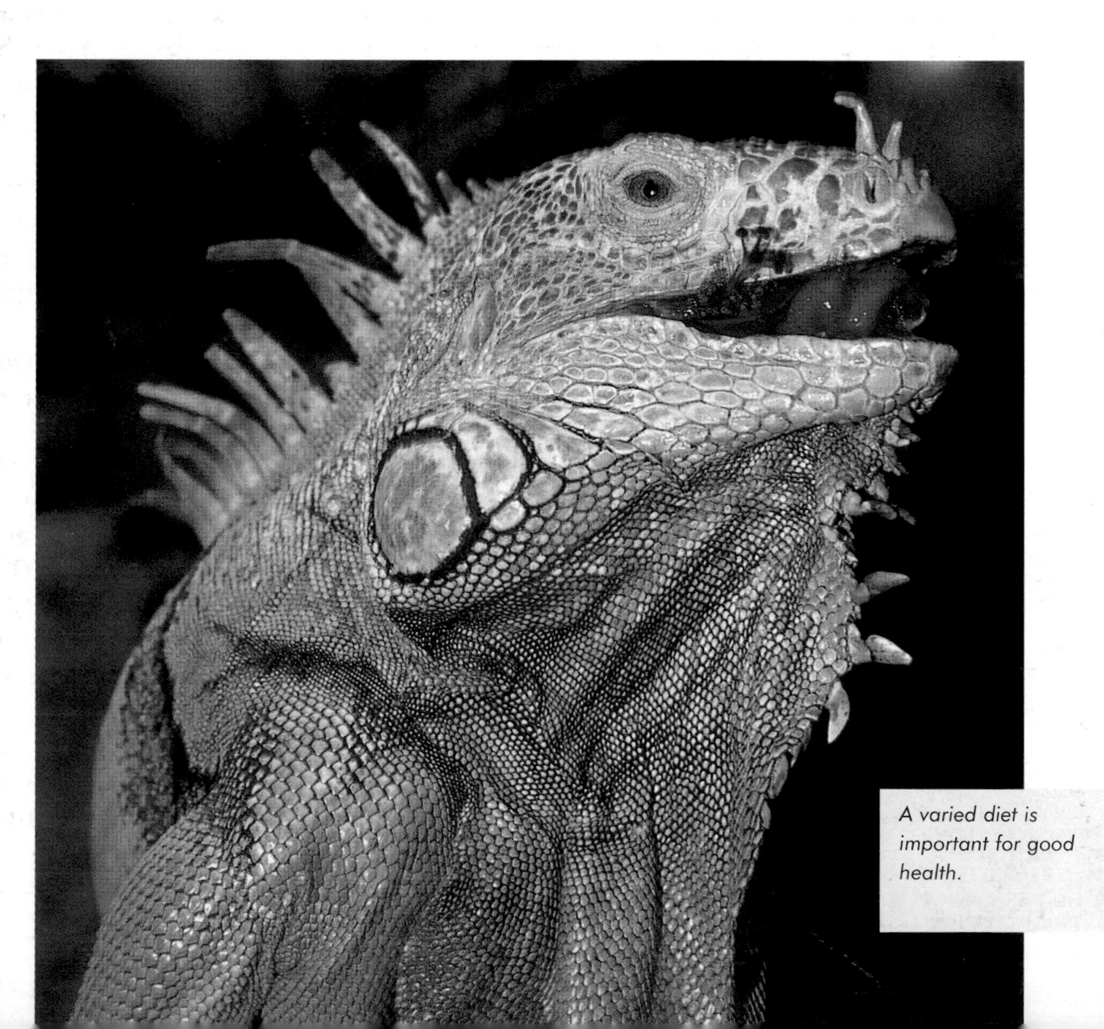

A varied diet is important for good health.

HERBIVOROUS DIETS

Amphibians will eat vegetable matter only when they are tadpoles.

The easiest group of creatures to feed are those which are primarily herbivorous. They can be offered a varied diet of greenstuff, vegetables and fruit, laced with a vitamin and mineral supplement, although stick insects generally will just eat bramble, or privet. Supposedly herbivorous reptiles may also eat invertebrates as well, although only on rare occasions. Tortoises roaming in the garden may seize upon a worm or snail for example and eat it, although they will not normally consume such creatures.

With increasing concerns about chemical sprays, it is possible to grow at least some food for herbivorous animals, such as cress indoors, or a wide range of plants in a garden. Avoid feeding large quantities of sharp-tasting vegetables such as spinach, which generally contain quite large amounts of oxalic acid, as this chemical interferes with the absorption of calcium from the gut. If you are growing this at home from seed, choose a low oxalic acid strain. It is also not a good idea to feed large amounts of cabbage, because this could depress the functioning of the thyroid glands and affect body metabolism.

A range of foods suitable for vegetarian reptiles.

Dandelions are very popular with most herbivorous reptiles. They can be collected in the wild, but only choose a site where weedkillers will not have been used. You can cultivate a supply of dandelions quite easily by cutting a long root into sections, and potting these up in moist soil. Kept in a sunny place, they will soon send up fresh shoots, enabling you to take a regular harvest of leaves. The flowers will also be eaten.

It is now possible to purchase complete foods for vegetarian lizards, especially iguanas. These are carefully formulated to provide all the necessary nutrients to keep these creatures in good condition. They are also convenient to use, and reasonably economical, since there is normally no need to spend extra money on dietary supplements. Canned food of this type tends to be more palatable than the dry kind. When transferring a lizard to this type of food, it will normally do no harm to mix a little greenstuff with the pellets, or even soak them, to improve their palatability, although any left uneaten will then have to be discarded before they turn moldy. It may not be a good idea to allow the lizards free access to such foods in any event, because they may become obese.

Cut all vegetable foods into pieces of suitable size and offer them in a bowl.

INSECTIVOROUS CREATURES

A wide selection of so-called "live foods" is now available from commercial suppliers, although in some cases, as with whiteworm (*Enchytraeus*) or fruit flies (*Drosophila*), you will need to acquire a starter kit, and culture your own regular supplies. These smaller invertebrates are ideal for young amphibians and spiderlings. As they grow older, hatchling crickets can be introduced to their diet.

A wide range of invertebrates can be fed to insectivorous creatures.

Crickets in fact represent one of the most versatile live foods, partly because they are available in a range of sizes. It is possible to purchase special vitamin and mineral dusting powders for these invertebrates, which should be sprinkled carefully over them before they are fed to the lizards or other creatures. This helps to correct nutritional deficiencies which could otherwise arise from the regular use of crickets as live foods. They are relatively low in calcium for example, and also deficient in vitamin A. If necessary, the crickets can be chilled in the fridge beforehand, to slow their level of activity and make them easier to handle. They can be kept until needed in a well covered container, with a damp sponge for moisture and grass for food.

Mealworms (*Tenebrio molitor*) are another staple live food that is available in various sizes, from small "mini-mealworms" upward. Unfortunately, their rather hard outer casing makes them indigestible for smaller species. There are now special foods available to improve their nutritional value, for those creatures which are able to digest them.

Mealworms should be kept in a sealed container with small ventilation holes, and can be fed on chicken meal (free from any drugs), with a slice of apple provided for moisture. In due course they will pupate, and emerge as small black beetles. It is possible to culture them in these surroundings, but it will be a slow process.

Other live foods are also available, depending to some extent on where you live. Wax moth larvae, for example, are an excellent conditioner for lizards, if fed alongside other items. Locusts, from small "hoppers" through to adults, can also be used as live food, especially for larger lizards such as monitors.

Specially cultured earthworms are also useful for a range of species, but garden worms need to be treated more cautiously, as they can spread parasites. Place them in a container lined with moist grass, so they can void the contents of their gut over the course of several days, before using them as live food.

It is of course possible to trawl through your garden for additional live food, – for example, spiders are generally a very popular food. A special deep entomological collecting net will be helpful for this purpose.

Crickets are widely available from specialist suppliers.

ANIMAL FOOD

It is possible to purchase a range of frozen mice, rats, chicks and similar foods from specialist reptile suppliers. These must be allowed to thaw out before being fed to snakes or lizards. Never be tempted to offer live animals to snakes. Not only may this be illegal, but a mouse or rat could turn and attack the snake in its vivarium. There is also a risk that the rodent might escape into the house. If necessary, forceps can be used to offer the prey, and encourage the snake to strike at its food.

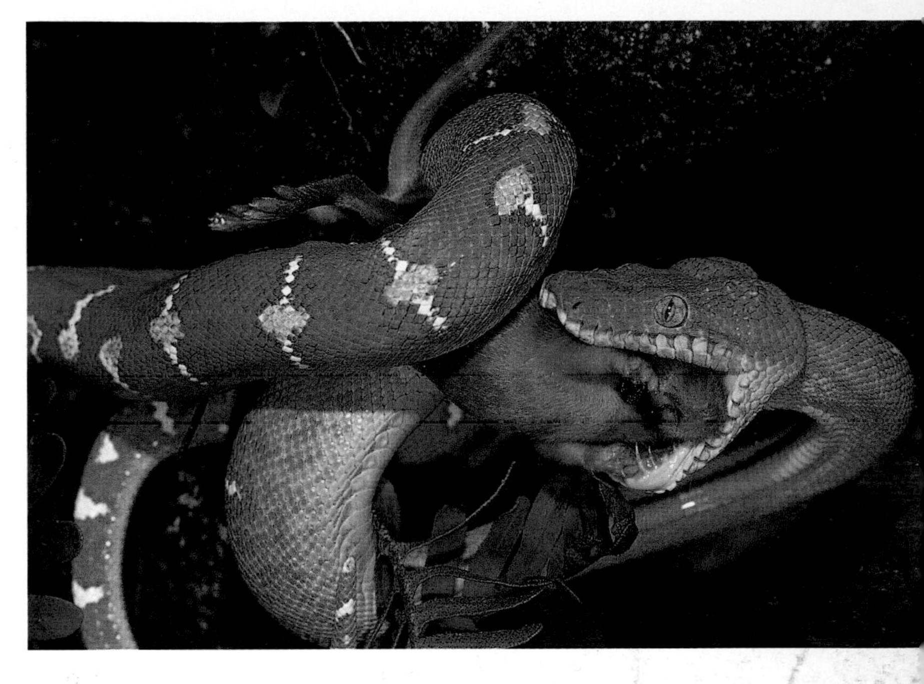

Food offered to snakes must be of a suitable size for them.

Mice and rats are sold by size, with "pinkies" being the smallest mice, killed at a day old, whereas so-called "fluffs" have a trace of fur on their bodies. The size of prey should correspond to that which the snake is used to eating, with feeding being carried out in accordance with the snake's appetite. It may take a week or so for a snake to regain its appetite after a move. Subsequently, it will probably need feeding once or twice a week, apart from insectivorous species which need daily feeding.

Dead rodents are much used as snake food.

As an alternative food, special sausages, formulated to contain all the essential nutrients for snakes, are now available in different sizes, corresponding to those of mice. These are a good alternative, particularly if you dislike handling rodents, and hatchlings in particular will take this type of food readily.

Fish will be eaten by some snakes and turtles, but there is a risk of thiamin deficiency arising from a diet of this type. When used as an occasional variant in the diet, however, fish can be beneficial.

SUPPLEMENTS

A wide range of vitamin and mineral supplements is now on the market, and these can be helpful, especially for recently imported reptiles. Care needs to be taken with the use of such products, however, because of the risk of overdosing, which can be fatal. Never be tempted to exceed the recommended dosage as stated on the packaging, nor use two products simultaneously. If in doubt, ask a veterinarian who is experienced with reptiles for advice.

Food supplements for reptiles and amphibians are available in powdered form.

HEALTH CARE

Sticky eyes will benefit from bathing. Reptiles will not eat if they cannot see.

Provided that you choose healthy stock that is feeding well at the outset, with luck you should not have serious problems. A reptile or amphibian that is in good condition will appear reasonably sleek, and should be able to move easily. Dragging of the hindlimbs is a particularly bad sign, frequently linked with metabolic bone disease in the case of lizards.

The signs of illness are usually fairly general, but in a few cases they can be quite specific. Turtles which swim at an abnormal angle in the water, with their body tilted, are likely to be suffering from respiratory disease affecting the lungs. Tortoises are also prone to this type of ailment, often manifested in this case by a discharge from the nostrils and eyes. Rapid treatment by a veterinarian represents the best hope of survival under these circumstances, and in all other cases where a creature is ill.

Unfortunately, it can be difficult to establish why a reptile or amphibian is losing weight. It may simply be that it has not been feeding properly, rather than being a sign of overt disease. Many lizards especially will soon recuperate if kept under an ultraviolet light, and rapidly put on weight.

Deworming can be a useful precaution, especially since parasites tend to become more prevalent when a lizard or other creature is out of condition. Mites can also be a problem in some cases, and may have a serious debilitating effect. Suitable treatments can be obtained from a veterinarian who specializes in the more exotic pets.

Ticks may also be seen occasionally, particularly on recently imported tortoises and snakes. They swell in size as they feed on blood and, like mites, they can also spread other diseases. Simply pulling the tick off is likely to leave its mouthparts anchored in the skin, where they could give rise to an infection. The best solution is therefore to smear the tick with petroleum jelly, blocking off the breathing pore at the rear of its body. This should persuade it to loosen its grip and it will drop off of its own accord.

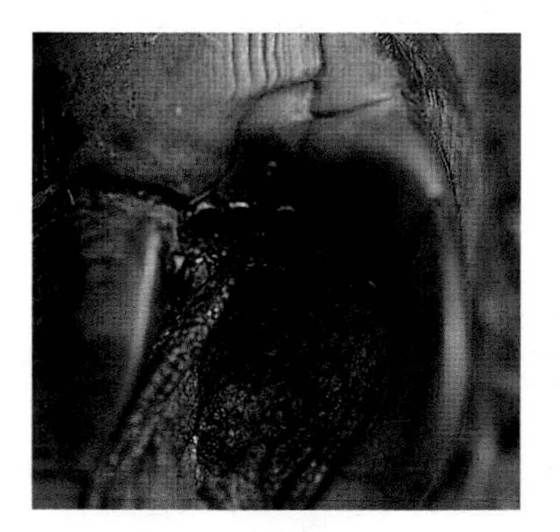

Parasites such as ticks are most likely to be seen on recently imported reptiles.

Even if you are purchasing a creature with a view to breeding from it in due course, always keep the newcomer housed on its own for a period of two weeks, to be sure that it is healthy and eating properly. Deworming and treatment for external parasites can also be carried out during this period. This should safeguard the health of your collection, for example preventing snake mites from becoming widely established even if you purchase a snake suffering from them.

FUNGAL DISEASE

Aquatic animals are especially susceptible to fungal infections. This can often be linked to poor water quality. Changing the water to remove the source of infection is essential, as well as treating the creature concerned. Although there are treatments that can be added to the water, these may damage the filtration system in some cases. Applying an ointment directly to the sick individual probably offers the best hope of a rapid recovery.

Fungal ailments often strike weakened individuals.

SALMONELLOSIS

This bacterial infection is a subject of concern among owners of reptiles because it is a zoonosis – a disease that can be spread from animals to people. It typically causes symptoms resembling severe food poisoning, and its effects can be especially grave in very young and old patients.

Reptiles can harbor the *Salmonella* bacteria in their intestinal tract without showing ill effects. It is possible to test for the presence of these bacteria by means of a fecal sample. Even if they are infected, however, good management should minimize the risk to human health. Washing your hands after touching any creature, or cleaning its quarters, is essential. It is also inadvisable to allow young children direct, unsupervised access to reptiles.

Dispose of contaminated litter from the vivarium carefully, and do not wash feeding dishes alongside food utensils for people. Dirty water should be tipped away down an outside drain, rather than a kitchen sink.

It is of course possible that the reptile could acquire *Salmonella* from its food. Chicken carcasses are often contaminated, for example, and so feeding raw meat of this kind could be dangerous. Specially prepared foods, such as the food sticks now available for turtles, are a much safer option.

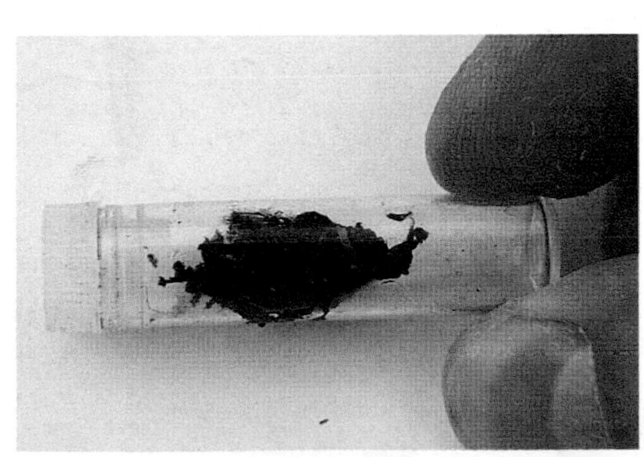

Suspected salmonellosis has to be confirmed by laboratory tests. Good hygiene will protect human health.

LUMPS AND BUMPS

The shells of tortoises and other chelonians can be damaged and bleed. They heal slowly.

Skin tumors do arise occasionally, and in the case of Lacertid lizards are often the result of a viral infection. Males acquire them on or around the face, and females are infected around the vent, as these tumors are spread by direct contact. This reflects the way in which the lizards court. It is possible to have these tumors removed surgically, but recurrences are common.

Not all lumps are necessarily tumors. In the case of chelonians, for example, swellings are often the result of abscesses. These will also need surgical treatment, especially if they are close to the head, as this may affect the reptile's ability to eat properly.

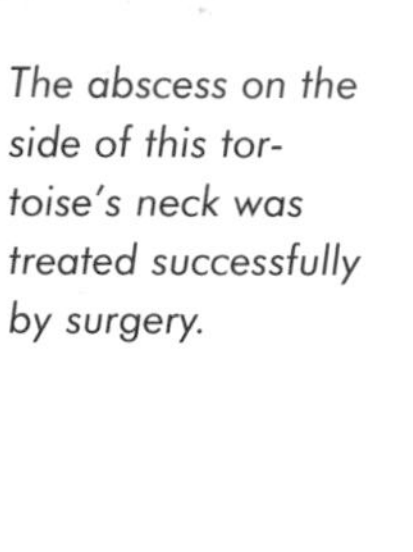

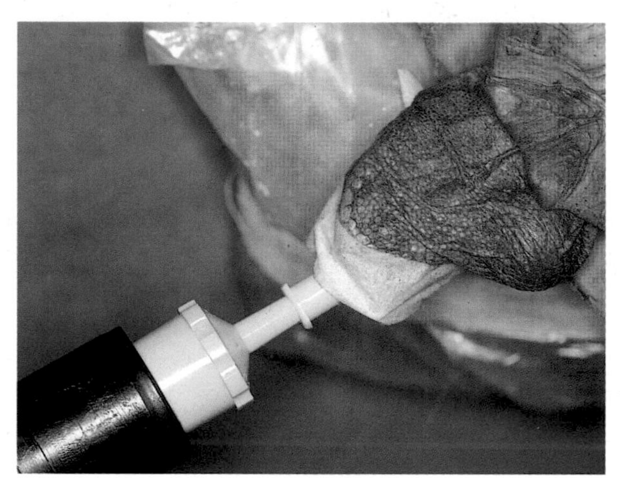

The abscess on the side of this tortoise's neck was treated successfully by surgery.

RED LEG

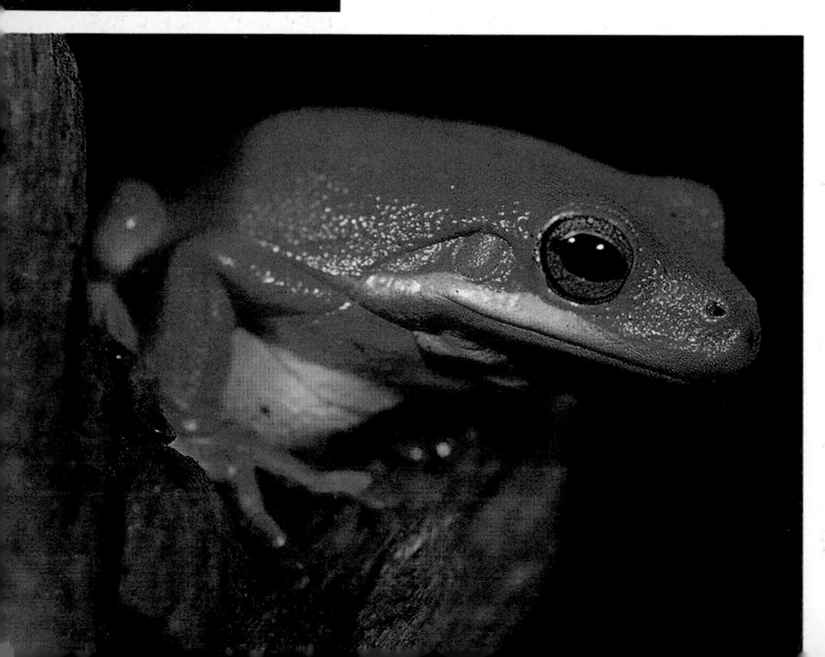

Although this is a healthy animal, skin injuries can allow bacteria to invade an amphibian's body.

This is a well documented infection of amphibians, particularly frogs. There is usually a characteristic reddening of the skin on the undersurface of the hindlegs, with death occurring soon afterward. Damage to the skin caused by rough handling could be a contributory factor in such cases. Rapid antibiotic treatment may sometimes lead to a successful recovery. Again, changing the water frequently, to lower the level of bacterial contamination here, may help to avoid the disease.

MOUTH ROT

This is another bacterial infection, which often strikes snakes in poor condition, and tortoises that have recently emerged from hibernation. The reptile loses interest in its food, and its mouth has an unpleasant odor. On closer examination, cheesy growths inside the mouth will be apparent. Your veterinarian may need to anesthetize the reptile in order to clean and treat the affected area properly. Antibiotics, and possibly the short-term use of a high-potency vitamin A supplement, may be useful.

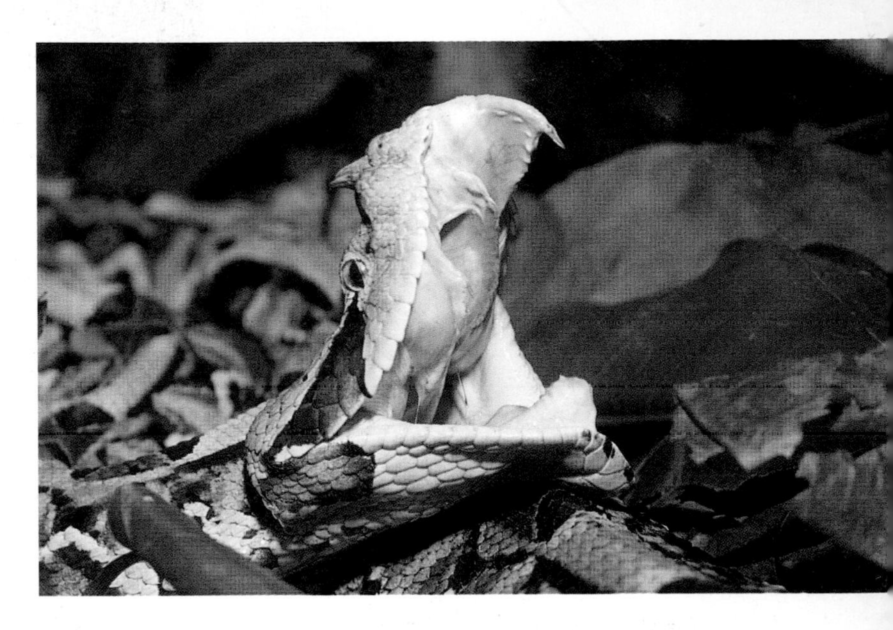

This snake is displaying a healthy mouth, but refusal to eat for some time can either be the result of, or the cause of, mouth rot.

VITAMIN AND MINERAL DEFICIENCY

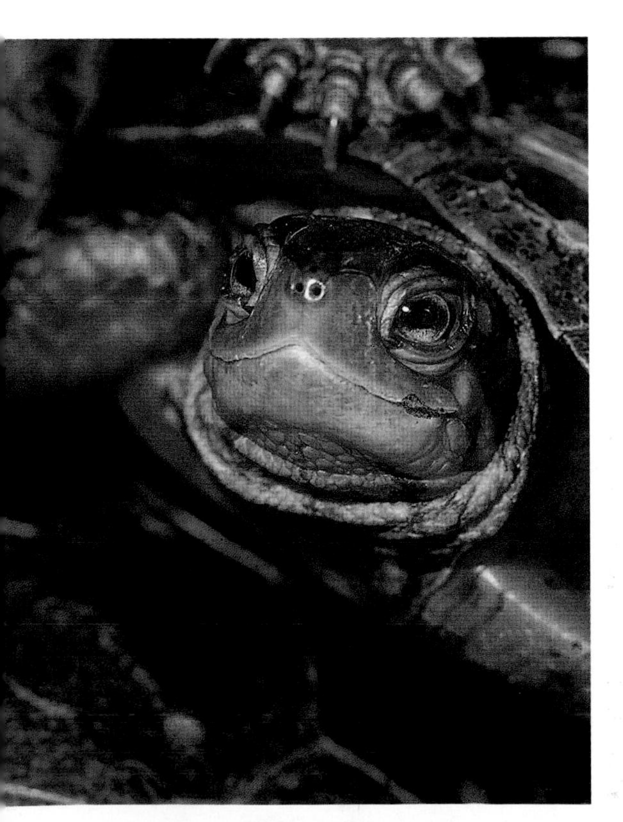

Bright eyes and a healthy shell, evident here in this spotted turtle, depend on correct vitamins and minerals in the diet.

Young terrapins are especially susceptible to vitamin A deficiency, which will arise if they are fed predominantly on a diet of meat. Their eyelids start to swell, and ultimately they will no longer be able to see, and so cease to feed. Internal changes leading to kidney damage occur, and ultimately prove fatal. Treatment in the early stages with vitamin A can reverse these signs. This condition is much less common than it used to be, as the special complete foods for these reptiles contain adequate vitamin A.

A deficiency of vitamin D_3 can also occur, causing weak bones, and soft shells in the case of chelonians. Proper lighting is essential to prevent these problems, as discussed earlier (see page 17). A lack of dietary calcium may also be implicated, however, and so supplementation of this mineral, especially in the diets of fast-growing herbivorous lizards, whose food is naturally deficient in this vital mineral, is recommended. Again, the risk is significantly lessened when the creatures are given a proprietary complete diet.

BREEDING

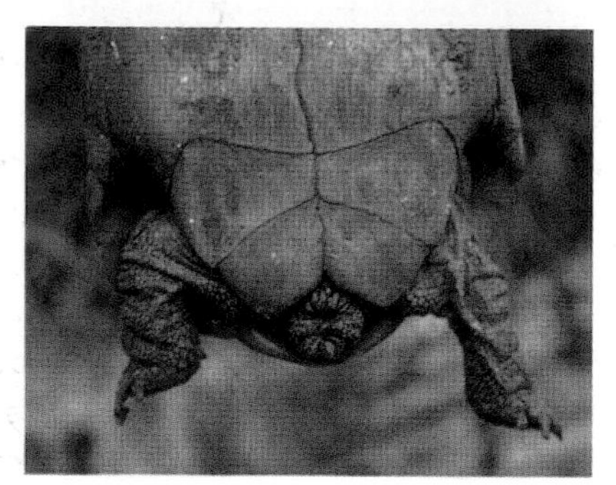

Chelonians may often be sexed by differences in tail length and shape. A male tortoise is shown above, a female below.

An improved understanding of the environmental and dietary needs of the creatures described in this book has made successful breeding commonplace for many species once thought to be difficult. Many snakes are now breeding readily, to the extent that color mutations are now well established in some species, creating a new area of study for those interested in genetics.

The first requirement in general for successful breeding is to be certain of obtaining pairs. In the case of snakes, they can be sexed if necessary by probing. The shape of the plastron and the relative length of the tail provide a useful pointer in chelonians. A curved plastron and a short tail usually indicates a female. Head embellishments such as crests, coupled with other less obvious indicators such as femoral pores running down the top of the leg, can help to reveal the sex of male lizards.

Newts can also be sexed by the size of their crests, which become pronounced in males at the start of the breeding period. The appearance of the cloacal region surrounding the vent can serve to indicate the sexes in the case of salamanders and newts. In either case it is more prominent in the male. Male frogs and toads often develop swellings known as nuptial pads on their forelegs, and they will become increasingly vocal as they come into breeding condition.

Sexing is also possible in some invertebrates, including tarantulas. Here males tend to be smaller than females, and have bulbous swellings, called palpal bulbs, on the legs at the front of their body. Females in those stick insects which do mate can generally be recognized by their larger size, and often brighter colors.

Successful breeding can be accomplished regularly if the environmental conditions are right.

The male tortoise needs the female's help when mating. If she does not stay still, he will not be able to balance properly.

THE BREEDING PERIOD

The breeding cycle in most species is closely allied with its environment, and the appropriate trigger, such as a fall in temperature, can set things in motion. Changes in behavior will often be the first sign that breeding is imminent.

Mating itself can be an aggressive encounter, with the male turtle for example often snapping violently at a female, in the hope of being able to grab on to her in the water and mate. Male frogs may also leap at a female ready to spawn, grabbing her in an embrace described as amplexus. Depending on the species, the male may grasp her either behind the front legs or lower down her body, with external fertilization taking place.

In the case of newts and salamanders however, the male releases a parcel of sperm, called a spermatophore, which the female takes into her body via her cloacal opening, thus fertilizing the eggs before they are laid.

Suitable conditions must be provided for egg laying to take place. In many cases, it will then be preferable to transfer the eggs, or any live young, to separate accommodation. There is no tightly defined incubation period for reptiles, unlike birds. The temperature required to hatch the eggs also tends to be lower, so that avian incubators must be turned down to their lowest setting.

Another highly significant difference is that reptile eggs should not be turned during incubation – this can kill the developing embryos. Indeed, they should be set in the incubator in the same position as they were collected in, if they have had to be removed from a substrate.

The incubator temperature is also significant when it comes to determining the sex of the hatchlings. Unlike mammals and birds, some reptiles, notably the crocodilians and chelonians, do not have sex chromosomes to determine their gender. This is determined by the incubation temperature, with a difference of just 2 °F(1 °C) likely to be critical. Studies suggest that there is no set pattern to this – it depends on the individual species. Some have female hatchlings at a higher temperature, others have males. There is an intermediate band where both sexes will be hatched.

A converted vivarium will suffice as an incubator if necessary, provided that the temperature here can be maintained around 84 °F (29 °C), with a relative humidity of 75–80%. Some eggs may show signs of mold on their shells during incubation. This is normal, and should not affect hatchability.

In due course, the young reptiles should crack through the shell with the sharp egg tooth on the bridge of the nose. This will then disappear, and the reptile will absorb the remains of its yolk sac into its body, before seeking food.

Amphibian eggs should be kept warm to speed up their hatching, because if kept too long they may develop fungus. It will soon be clear which are fertile, as shown by the development of tadpoles in the semi-transparent jelly. Suitable weaning foods will be required once they hatch, and the tadpoles should then be separated into smaller batches of similar size, to prevent any cannibalism.

The same applies in the case of young spiderlings when they hatch from their communal egg case. Almost inevitably under such circumstances, some will be eaten by their siblings, as would occur in the wild.

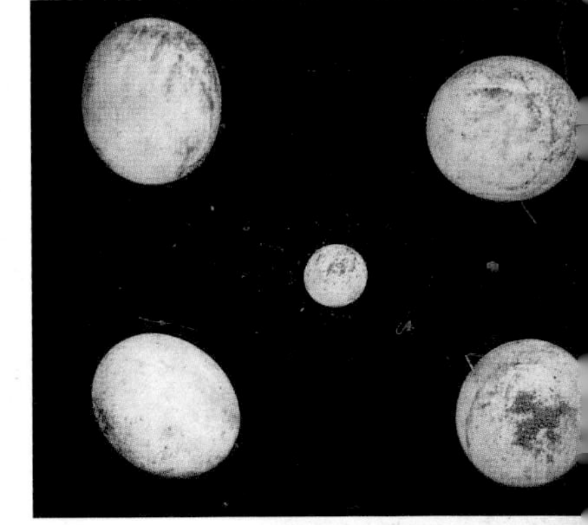

Eggs need to be transferred carefully to an incubator without being turned. Very small eggs will not hatch.

CONSERVATION CONTROLS

ENDANGERED SPECIES

It may be illegal to catch certain reptiles and amphibians in the wild, even if you are simply intending to move them to a pond on your property for example. In places where it is illegal to capture amphibians and reptiles, it may be illegal to own such species, even if you did not catch them yourself. You must check up on any laws that may apply, because law enforcement agencies do not consider ignorance of the law an acceptable defense.

When purchasing stock, buy only from reputable dealers, and beware of any surprisingly cheap offers, particularly of rare species. The herptiles in question might be ill, or they may have been smuggled into the country illegally. The Convention of International Trade in Endangered Species (best known under its acronym of CITES, pronounced si-tees) regulates the international movement of wild fauna and flora, including herptiles.

Even common species, such as this timber rattle snake can be endangered even though it may not be the whole species that is threatened with extinction, but rather a particular population in a specific area.

It operates primarily by means of listings of species in appendices. Those creatures or plants that are deemed to be endangered are featured in Appendix I, which outlaws any international commercial trade in wild caught specimens, although zoos or other recognized breeding institutes may be able to obtain stock for captive-breeding purposes having gained the necessary official approval and permits.

Species that are considered vulnerable are featured in Appendix II, which allows for monitoring of trade levels. Should these give rise for concern, then a variety of measures may be instituted, from the imposition of quotas to a temporary cessation of trade, pending an assessment of more detailed population data.

The tendency now is to set quotas for Appendix II species in any event, based on field studies, so that there should be no risk of a population becoming over-exploited. This type of wildlife management is known as sustainable use (SU). It allows local people to benefit from their wildlife resources, instead of clearing tropical forest for agricultural purposes, for example, and so helps to conserve the natural environment. Trade can be a very positive conservation tool in this respect.

Import and export permits are likely to be required for international movement of CITES Appendix II species, and this paperwork will need to be obtained in advance, even if you are simply moving a small number of specimens from one country to another. Captive-bred herptiles may also fall under the CITES regulations.

For advice on the current legislation, listings, or other relevant matters including transportation requirements, you should contact your CITES Management Authority, which is responsible for the implementation of the convention. This is likely to be the government department concerned with nature conservation in your country, and so should not be too difficult to track down.

Countries which are so-called "Parties" or members of CITES (there are now more than 120 "Parties") meet roughly every 2½ years, and decide on proposed changes to the listings in the Appendices. Species can be moved up and down, depending on the latest information. In between this major conference of the Parties, various groups, notably the CITES Animals Committee, meet regularly to continue the review process. The headquarters of CITES is in Switzerland, where a full-time Secretariat oversees its operation.

The more colorful species, such as the milk snake, are in particular demand, so that controls on trade in wild-caught specimens are often necessary.

DANGEROUS ANIMALS

Size is not necessarily an indication of danger. Poison arrow frogs have long been used by native people in parts of Central and South America as a source of poison to tip darts for hunting.

Tomato frogs – bright coloration in the animal kingdom is frequently a sign of toxicity, warning predators not to attack.

It pays to be sensible with both household and exotic pets. While a dog or cat may defend itself by biting, some exotic animals produce poisons. Many amphibians have toxic skin secretions, with their attractive bright coloration often serving to warn potential predators of this threat.

These are unlikely to be a hazard for people, particularly as members of this group of creatures rarely have to be handled directly. On those occasions when this is necessary, a pair of disposable gloves will prevent any toxin from gaining access to the body via cuts or abrasions on your hands.

Unfortunately, other herptiles are likely to be aggressive. The keeping of venomous snakes is a controversial area, not only because of the potential risk to the handler, but also because of fears that such snakes could escape and present a serious danger to the neighborhood.

It is not just the bite of a large spider that can have unpleasant effects. Its hairy body can irritate the skin, so handle with care.

As a result, in many countries, there are strict controls in place, including registration of owners of species that are thought to represent a serious potential threat to people. The facilities for poisonous snakes will also probably have to be checked, particularly with regard to their security.

Other reptiles may also prove to be harmful if mishandled, such as large constrictor snakes, which should never to allowed to wrap themselves around children. Crocodilians (embracing crocodiles, alligators, and caimans) are perceived to be especially dangerous because of their strength. Large crocodilians do kill people in their homelands each year, although in fairness, only about six of their number represent a serious danger in this respect. Nevertheless, with a mouthful of sharp teeth, any bite even from a small crocodilian will be painful. The nature of the bite also usually means that it is slow to heal, with the points of the teeth penetrating deep within the flesh.

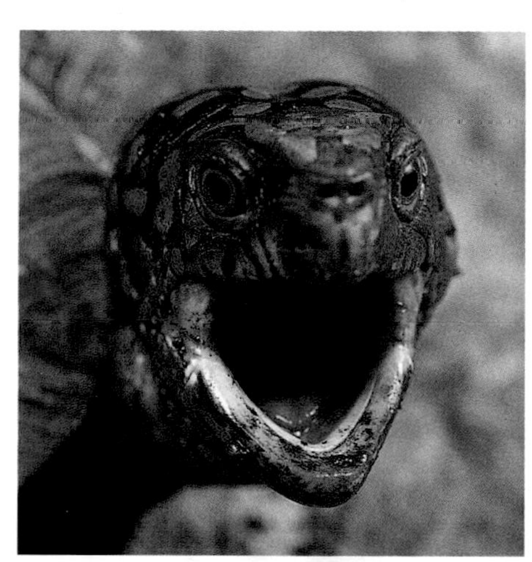

Even a turtle may inadvertently bite your finger if it is in the wrong place when you are feeding the creature.

In spite of their often rather sluggish appearance, crocodilians can move very fast over short distances, grabbing their prey with ferocity. Even small individuals can turn and snap with amazing alacrity if threatened. Although caimans are sometimes available as hatchlings, they therefore cannot be recommended as pets. Their care is demanding, quite apart from their potential to cause injury.

Regulations governing the sale of potentially dangerous pets are therefore also applied to crocodilians in various countries, with a special permit required to keep them. There may be other stipulations as well, such as compulsory third-party insurance.

Information on the requirements for keeping dangerous species can usually be obtained from local government offices, and again, the necessary advice should be sought well in advance of obtaining the species in question. These herptiles are very much for the specialist and cannot be recommended for the novice owner. With so many other species being regularly kept and bred, there is really no need to become involved in this field, at least until you have gained considerable experience with non-poisonous snakes, for example, and are certain that you want to take such a step, being fully aware of all the implications.

The creatures featured in this part of the book represent a wide selection of the species that are often available and widely kept, although not all are suitable for the novice owner. Each entry is standardized, with the difficulty of the care of each species being coded on a scale of 1 to 10. Higher figures indicate that a greater degree of experience or commitment will be required – possibly, for example, because of the difficulties in housing a particular species.

The sizes given in these charts indicate a figure from the upper range of the scale in each case, allowing for individual variations, and relate to the creature's overall length. Lizards for example are sometimes measured on the basis of their snout-vent distance – the so-called "s-v" measurement – since part of the tail may be missing. This figure is of little practical value in terms of assessing the housing needs of a particular species, however, because tail length may be equal to or might even exceed the s-v measurement figure.

The basic type of environment required by each of the creatures is indicated, along with its distribution in the wild. An assessment of compatibility is also included – although individuals may differ in their behavior, depending on whether they are in breeding condition, for example, when they are more likely to become aggressive. Temperature requirements can be found in the relevant pages for each housing type in the first part of the book.

The type of food required is set out here in general terms, but some cautious experiment can be helpful. Where amphibians are described as carnivorous, this refers to worms, slugs and similar invertebrates that are their preferred prey, rather than young mice or meat.

Breeding data is also included, but it should be remembered that young animals breeding for the first time are unlikely to produce as many eggs or offspring as those that are fully mature.

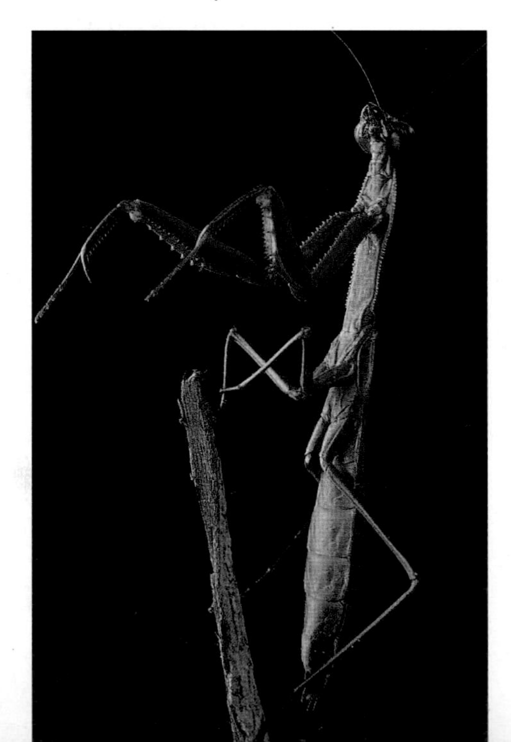

The individual needs of the creatures covered here, such as the praying mantis (left) and the agama (above), can be quite specific.

HOW TO USE THE SYMBOLS

A unique feature of the directory is the set of symbols which accompany each entry. This provides key information at a glance, covering topics such as the likely adult size, habitat, housing, temperament, feeding requirements, reproductive data and ease of care. Since the care of creatures even within the same category may vary widely, such an approach will help to guide the novice owner, particularly when deciding which species would be the most suitable choice of pet.

SYMBOLS

Amphibians

1. Non-aggressive or social nature
2. Predatory by nature, especially with smaller companions

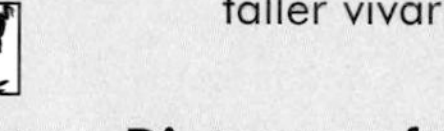

3. Aggressive and anti-social
4. Likely number of eggs or young

Invertebrates

5. Non-aggressive or social nature
6. Predatory by nature, especially with smaller companions
7. Aggressive and anti-social
8. Likely number of eggs or young

Reptiles

9. Non-aggressive or social nature
10. Predatory by nature, especially with smaller companions

11. Aggressive and anti-social
12. Likely number of eggs or young

Vivarium design

13. Essentially terrestrial by nature
14. Arboreal, requiring taller vivarium

Dietary preferences

15. Primarily carnivorous
16. Primarily insectivorous
17. Primarily vegetarian
18. Eats plant and animal foods readily

Vivarium type

19. Savannah
20. Semi-aquatic
21. Temperate woodland
22. Aquatic
23. Tropical woodland
24. Desert

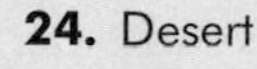

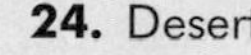

Size

25. Likely adult size, in imperial and metric units.

Ease of care

These symbols, graded from 1–10, provide a means of comparing the needs of the different creatures. The ratings give an objective practical assessment of the overall difficulty in maintaining a species. Figures at the low end of the scale indicate the species most suitable for beginners while those close to 10 call for experience. Their housing or feeding needs, for example, may be more demanding than in the case of other species.

REPTILES

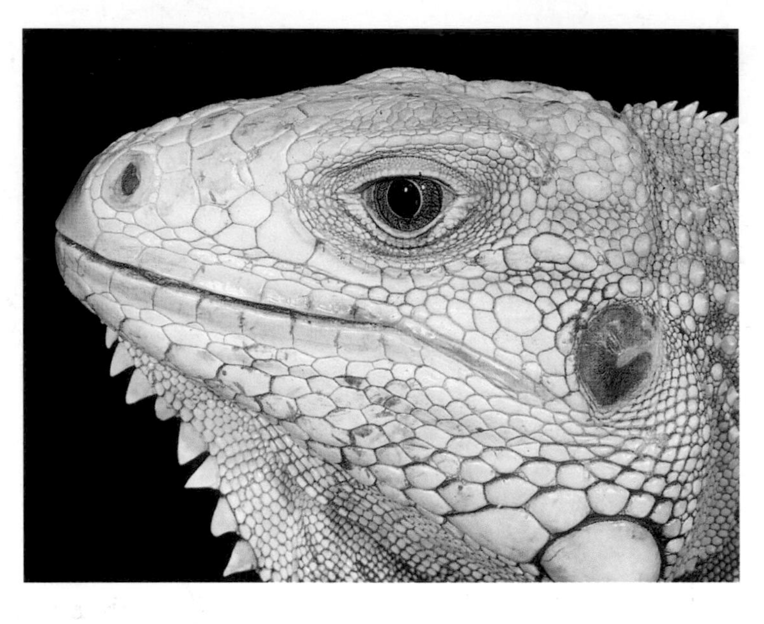

The scalation on the head of this green iguana is clearly visible. The dewlap skin – under the chin – is inflatable in some lizards, and is used for display.

Snakes and lizards are quite closely related, with the rudiments of legs being present as spurs in some snakes, such as the boa constrictor. On the other hand some lizards, notably the slow-worm (*Anguis fragilis*) have no legs and resemble snakes.

Handling these two groups of reptiles successfully depends partly on the size of the individual, and clearly on whether or not it is poisonous. If it is, special protection and tools are advisable. Some larger reptiles can be aggressive and will bite, as well as attempting to lash out with their powerful tails and sharp claws.

It is important to bear in mind that many lizards will shed their tails readily if badly handled, in the same way as if they were seized by a predator in the wild. They need to be restrained firmly by the head, with the tail tucked into the side of the body, to prevent them from struggling. Non-poisonous snakes require similar handling, being restrained behind the head, with their body looped around the hands for support.

HANDLING A LIZARD

Place your finger and thumb each side of the head to keep a lizard from struggling when picked up. Watch out for the claws, which can be sharp, and never try to restrain these reptiles by the tip of their tail.

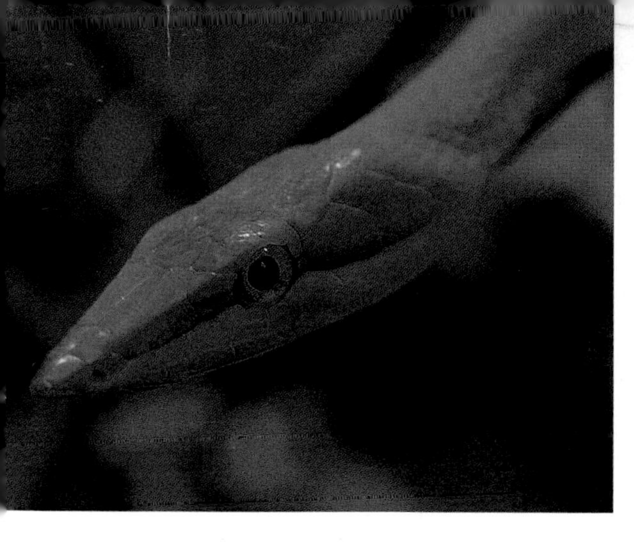

Venomous snakes, such as this green vine snake, should only be handled by experts.

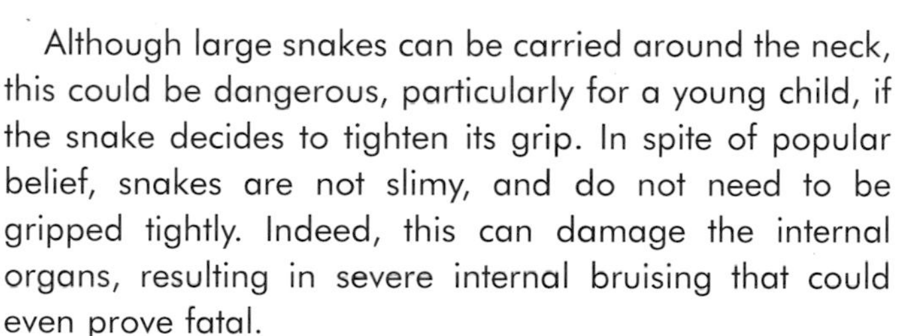

HANDLING A SNAKE
Place one hand behind the snake's head when picking it up. You can then hold the snake quite easily by gently wrapping your fingers around its upper parts and using your other hand to support the rest of the body.

Although large snakes can be carried around the neck, this could be dangerous, particularly for a young child, if the snake decides to tighten its grip. In spite of popular belief, snakes are not slimy, and do not need to be gripped tightly. Indeed, this can damage the internal organs, resulting in severe internal bruising that could even prove fatal.

The chelonians – tortoises, terrapins and turtles – can usually be handled without difficulty simply by picking them up by their shells. Some will struggle with their legs however, which could cause you to drop them, possibly damaging their shells.

Be sure to have a firm grip on both the top and bottom of the shell on the sides when handling these particular reptiles, and in the case of those with hinged lower plastrons, be careful that you do not get a finger trapped when the reptile tries to retreat fully back into its shell as this could be painful. Aquatic species can also prove to be slippery, when their shells are wet. Snapping turtles and terrapins can inflict serious bites.

Chameleons anchor themselves firmly to a branch so lift their feet off carefully.

HANDLING A TURTLE
Most chelonians withdraw into their shells when picked up, but if held for any length of time, they may start struggling.

20 in. (50 cm)

Tropical woodland

Terrestrial

Aggressive

Insectivorous

2–8 eggs

AMEIVA

Ameiva ameiva

Originating from Central and South America, these lively lizards are sometimes called jungle runners, because of the speed at which they can move if disturbed. Ameivas are territorial by nature, and pairs should be kept apart. The markings of individuals may vary quite widely, and sexing on the basis of their coloration alone is therefore difficult, although females tend to be less vividly marked than males. Other pointers which can help to distinguish the sexes are that females are often smaller and have less powerful jaws.

A good source of ultraviolet lighting is essential for ameivas to keep them in good health, and the substrate of their vivarium should allow for their tendency to dig regularly. Although they are mainly insectivorous in their feeding habits, ameivas will eat a little fruit.

REPRODUCTION Pairs may breed successfully in vivarium surroundings, with the eggs buried beneath a low mound created by the female lizard. It is not a good idea to leave the nest intact here, however, because ameivas have a tendency to eat their own eggs. They may produce two clutches of eggs each year, with hatching likely to occur from about 6 weeks on.

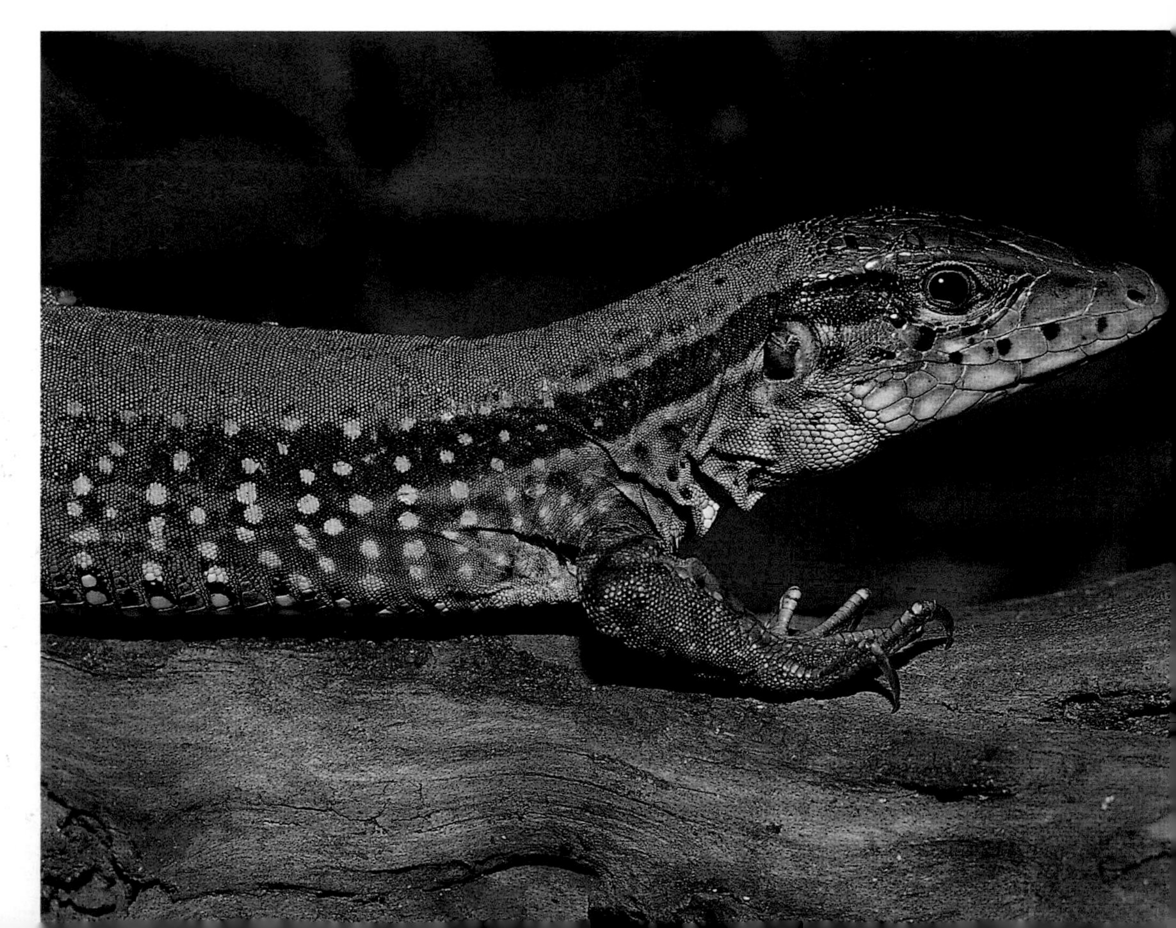

AMEIVA
Ameivas are lizards to look at, rather than handle, in view of their nervous and lively nature.

SLOWWORM

Anguis fragilis

This species is found in Europe, Asia, and North Africa. Although sometimes thought to be a snake and not a lizard, the slowworm has movable eyelids, which allow it to blink. Slowworms also shed their tail if threatened, like many other lizards. The tip will grow back, but it may not reach its former length and tends to have a stubby rather than a pointed end.

Slowworms make fascinating vivarium occupants and are the longest lived of all lizards.

16 in. (40 cm)

Temperate woodland

Terrestrial

Agree in groups

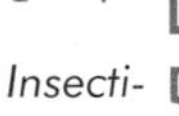
Insecti-vorous

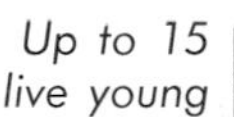
Up to 15 live young

REPRODUCTION Mating will occur in early spring, with the male grasping the female and entwining around her. The young are born approximately 3 months later, breaking out of the egg cases as they emerge from their mother's body. They measure about 3 in. (8 cm), and are gray in color, with a dark band down the center of their backs. Sexual maturity is reached by the time they have grown to approximately 10 in. (25 cm) in length.

SLOWWORM
Female slowworms are of a more coppery color than males, with a dark streak down their backs.

SOUTHERN ALLIGATOR LIZARD

Gerrhonotus multicarinatus

These lizards come from the west coast of North America. The name of "alligator" lizard refers to the large, raised body scales, which are like those of alligators. They are not difficult lizards to keep in a vivarium, being relatively hardy and spending most of their time on the ground. A large floor area, and a few branches for climbing, should be incorporated into the design. There also needs to be a shallow container of water, since these lizards inhabit reasonably moist areas in the wild.

20 in. (50 cm)

Semi-aquatic

Terrestrial

Aggressive

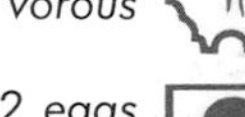
Insecti-vorous

6–12 eggs

REPRODUCTION This species produces eggs, but those in upland areas produce live young. Incubation lasts about 6 weeks, with the eggs being incubated at a temperature of 82°F (28°C). Young alligator lizards measure approximately 4 in. (10 cm) in length on hatching.

SOUTHERN ALLIGATOR LIZARD
The base of a male's tail is usually thicker than a female's.

ASIAN WATER DRAGON

Physignathus cocincinus

Young hatchling water dragons are cute creatures, but it is important to remember that they will become much larger and need spacious surroundings as they grow. The ideal type of environment for mature water dragons is an enclosure equipped with a pool, rather than a vivarium. These lizards are found in India and Southeast Asia. They live close to water and are excellent swimmers. They are also agile climbers and can run up a tree on their long hind legs if they are threatened on the ground.

It is much better to start out with young water dragons, because they are invariably quite tame, whereas adults usually prove to be wild and will be most unlikely to calm down. The banded patterning typically seen in juveniles will disappear as they mature, and a crest will develop on their backs, extending from the neck down to the tip of the tail. Males can be recognized by their larger crests and bigger overall size. It is better to keep them apart, unless their surroundings are very large, because they may fight.

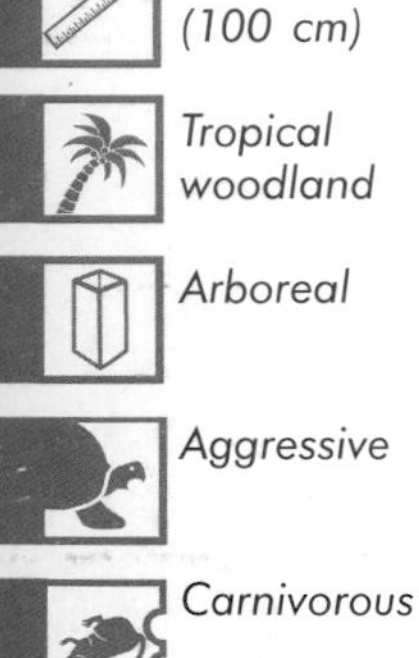

40 in. (100 cm)

Tropical woodland

Arboreal

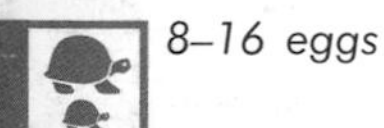

Aggressive

Carnivorous

8–16 eggs

REPRODUCTION **Female water dragons may lay several clutches of eggs, usually during winter and early spring. A suitable container, such as a plastic baby bathtub, filled with damp sand will be required by the female for egg laying. The eggs should start to hatch after 2 months, when incubated at a temperature of 84°F (29°C), and the young water dragons will be about 6 in. (15 cm) long when they emerge.**

Rearing is reasonably straightforward, but the young can be susceptible to limb weaknesses, which are usually the result of dietary deficiencies. A varied diet, including pinkies rather than just invertebrates, plus suitable exposure to ultraviolet light, to maintain their vitamin D_3 level, will be essential. Water dragons grow rapidly, reaching nearly 16 in. (40 cm) after 1 year.

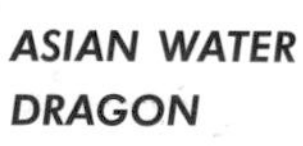

ASIAN WATER DRAGON
Provide plenty of climbing opportunities for Asian water dragons.

PANTHER CHAMELEON

Chamaeleo pardalis

PANTHER CHAMELEON
Males are very territorial and will turn a vivid red and yellow when threatened.

12 in. (30 cm)

Tropical woodland

Arboreal

Aggressive

Insecti-vorous

30–50 eggs

These chameleons come from Madagascar. They have feet that are specially adapted so they can wander along thin branches without losing their grip. They are also able to capture their prey at a speed faster than the human eye can register. The sticky tip of the chameleon's tongue allows it to seize flies and similar tiny creatures, which are then pulled back almost instantaneously into its mouth.

When resting, the male can be recognized by its predominantly dark green body with light blue stripes.

Unfortunately, chameleons are not easy to keep as pets, although advances in their care have been made in recent years.

REPRODUCTION Females, recognizable by their shorter snout, indicate their readiness to mate by becoming a shade of salmon orange. Later their color will darken. These chameleons produce eggs that are buried in the ground. Incubation may last from 240 to 300 days. When the young emerge, they are like miniature adults, and will climb and hunt almost immediately.

JACKSON'S CHAMELEON

Chamaeleo jacksoni

14 in. (35 cm)

Temperate woodland

Arboreal

Aggressive

Insectivorous

Up to 30 live young

This chameleon lives at fairly high altitudes in East Africa and can survive at temperatures as low as 40°F (4°C). Jackson's chameleons can even be kept outside in well-planted enclosures in mild climates. Some variation in the temperature of their environment appears to be beneficial.

Chameleons rarely drink from a container so, especially with recently imported individuals, use a dropper to place droplets of water (augmented with a vitamin and mineral supplement) on their lips. Once they are established, spraying the vegetation in their enclosure daily should provide them with adequate fluid intake. It is often a good idea to treat newly acquired chameleons under veterinary supervision for tapeworms, as their feeding habits appear to make them vulnerable to these parasites. Chameleons rank among the shortest-lived of all lizards, so purchasing youngsters rather than adults of indeterminate age is definitely to be recommended. A varied diet of live food is essential to keep them healthy.

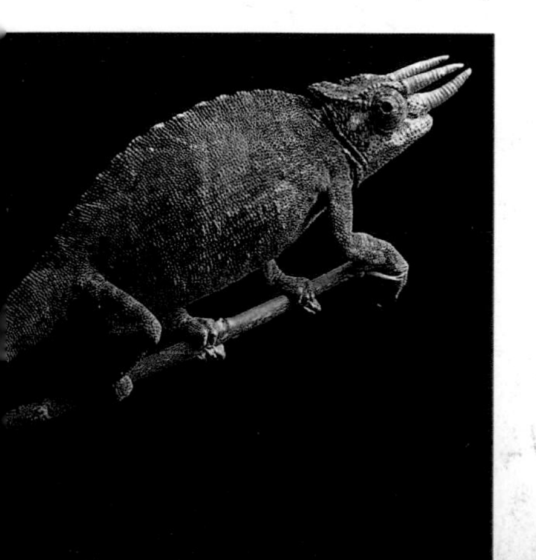

JACKSON'S CHAMELEON
Chameleons will benefit from being kept in spacious enclosures.

REPRODUCTION Sexing these chameleons is straightforward, because only males have the three long "horns" that have given rise to the species' alternative name of three-horned chameleon. As with other chameleons, males should be kept apart from each other, because they will suffer stress over a period of time from constantly challenging each other. Jackson's chameleon is one of the easiest species to breed, with females giving birth to as many as 30 young.

GIANT DAY GECKO

10 in. (25 cm)

Tropical woodland

Arboreal

Aggressive

Insectivorous

Two eggs

Phelsuma madagascarensis

The brilliant coloration of these lizards, and the ease with which they can be kept and bred, has ensured their popularity. In contrast to geckos which are more active after dark, day geckos in general have round rather than slit-shaped pupils. Their toes have broad pads that allow them to climb vertical surfaces easily. The giant day gecko is the largest member of the group and originates from the island of Madagascar.

A diet of small invertebrates will suit these lizards well, but other items, such as a nectar solution (as sold for birds) or honey dissolved in water should be provided fresh each day. Their enclosure should be well planted, as males can be quarrelsome, even toward their intended mates.

REPRODUCTION **Sexing is straightforward, with the presence of large femoral pores indicating a male. Females lay clutches of two eggs, concealing them in hollows. Bamboo tubes are often used for this purpose. Maintained in separate accommodation, at a temperature of about 82°F (28°C), the young geckos should emerge between 2 and 3 months later. As with adults, ultraviolet light is essential to their well-being.**

GIANT DAY GECKO
Day geckos can be prolific breeders, laying repeatedly through the year.

LEOPARD GECKO

10 in. (25 cm)

Desert

Terrestrial

Aggressive

Insectivorous

Two eggs

Eublepharis macularius

This species is ideal for those starting out with lizards. They are easy to look after, as well as hardy. Leopard geckos are found in western Asia, as far as Northwest India. They benefit from a considerable variation between their day and nighttime temperatures, as occurs in their natural habitat. The temperature can be allowed to rise to nearly 104°F (40°C) in part of the vivarium during the day, with the interior cooling down to 68°F (20°C) overnight. Rocks or other tank decor to provide retreats, and a shallow bowl of water, are essential. A period of winter dormancy is also beneficial, especially for breeding.

REPRODUCTION **The female will lay on a damp patch of the substrate, often close to a water bowl, depositing 2 or occasionally 3 relatively soft-shelled eggs here. Again, it is safer to remove the eggs for incubation purposes. Hatching is likely to take between 6 and 8 weeks on average, at a temperature up to 86°F (30°C). Young leopard geckos have brownish and pale yellow banding, rather than the mottled appearance of adults.**

LEOPARD GECKO
The leopard gecko lacks the adhesive footpads characteristic of many geckos.

TOKAY GECKO

Gekko gecko

Although these geckos are easy to keep, they are not especially friendly; they may attempt to bite when being handled, so gloves are recommended. They are found in Southeast Asia and on islands up to the far west of New Guinea. Their name comes from the sound of the call which males utter frequently. They should only be kept in true pairs, and retreats within their enclosure are important. Avoid placing tokay geckos of widely differing sizes together, because there is a risk that the smaller individual could be attacked or even eaten by its companion. It is equally important to shield any suspended heat source within the vivarium, because the agile nature of these geckos means they could climb on it and get burned. The type of food offered should reflect the size of the gecko. Young tokays should be provided with small invertebrates, larger individuals are able to consume pinkies, which provide a more balanced diet.

REPRODUCTION Sexing these geckos is straightforward, with males showing large femoral pores. These can be seen quite easily if the gecko is transferred to a clear plastic-bottomed container. The swelling near the base of the tail is another indicator of a male.

A calcium supplement is recommended for these geckos, which lay calcareous eggs. This can take the form of scrapings from a cuttlefish bone or a calcium supplement (as sold for pet birds). Females will store calcium in special neck glands, which can create the impression of swellings here.

The eggs are sticky and are likely to be hidden near to the roof of the vivarium. Providing a hollow tube for egg laying allows the eggs to be taken out easily; otherwise, they may be stuck to the sides of the vivarium and cannot be removed safely without risk of damage. The incubation period is extremely variable, being likely to last between 17 and 29 weeks. Young tokay geckos measure about 4 in. (10 cm) in length when they emerge from their eggs.

14 in. (35 cm)

Tropical woodland

Arboreal

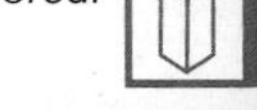

Aggressive

Insectivorous

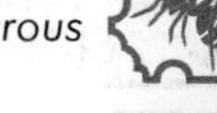

2 eggs

TOKAY GECKO
Orange and blue spots on a grayish-blue background characterize the tokay gecko.

20 in.
(50 cm)

Savannah

Terrestrial

Aggressive

Vegetarian

6–25 live young

EASTERN BLUE-TONGUED SKINK

Tiliqua scincoides

These large skinks are found in New Guinea, and northern and eastern Australia. They rank among the most expensive lizards. They are not difficult to keep and can become tame, particularly when handled regularly from an early age. Eastern blue-tongued skinks can be identified by the 7 to 10 distinct dark bands running across the back of their bodies. A dark area extends back from the snout through the eyes. The most remarkable feature of this lizard is its blue tongue, which it will display while hissing if disturbed. Their legs are short, and generally they move slowly, although they can speed up over short distances. While mainly vegetarian, they will eat various invertebrates, especially snails.

REPRODUCTION **Mature males tend to have larger heads than females. They will pursue intended mates around their quarters until the female allows the male to balance on top of her, grasping her neck with his jaws. Larger females tend to have bigger broods, with the young born about 4 months after mating. Rearing is reasonably easy; small invertebrates feature in the diet.**

BLUE-TONGUED SKINK
Blue-tongued skinks like to bathe, so provide them with a suitable water bowl.

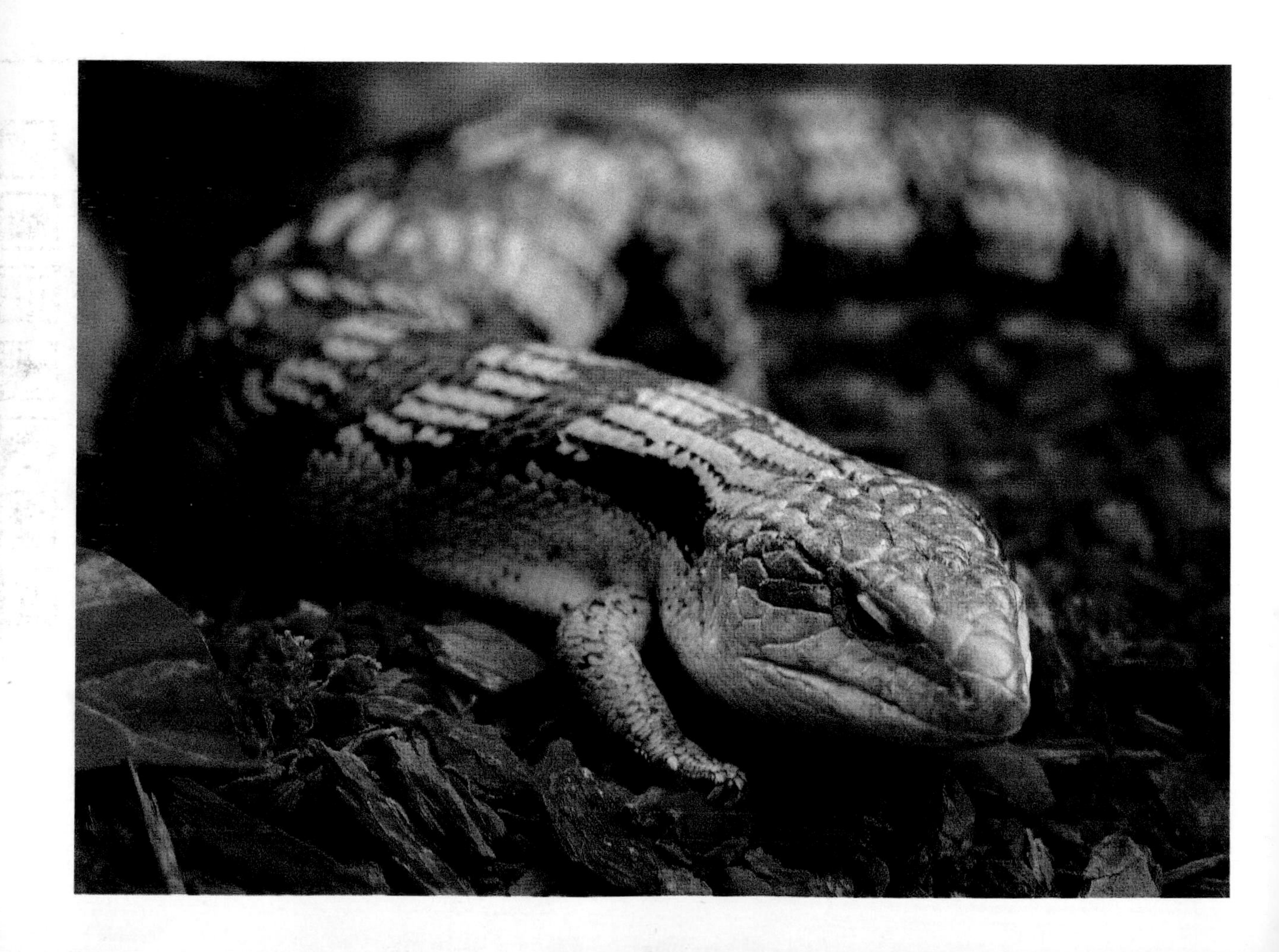

COMMON WALL LIZARD

Podarcis muralis

8 in. (20 cm)

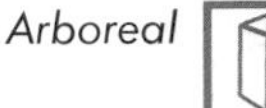

Savannah

Arboreal

Aggressive

Insectivorous

3–8 eggs

This tends to be the most commonly kept species, but there are other wall lizards which may be available, and all require similar care. They are found in southern and central Europe and, as their name suggests, they are often found in rocky areas and sun themselves on walls, darting off if danger threatens. Their markings may vary quite widely, even between members of the same species, and occasionally some melanistic individuals, which are predominantly black in color, may be encountered.

REPRODUCTION Mating normally begins several weeks after the end of the winter period, when the temperature in their quarters is raised again. The male starts to nudge the female, and when she is receptive, the female will remain in position, allowing mating to take place. Subsequently, after an interval of 10 days or so, the eggs will be laid in a damp locality in the vivarium. The young wall lizards should then emerge just over 6 weeks later. They can be reared in groups at this stage, requiring a varied diet and exposure to ultraviolet light.

COMMON WALL LIZARD
Female wall lizards may lay three clutches in a year.

GREEN LIZARD

Lacerta viridis

16 in. (40 cm)

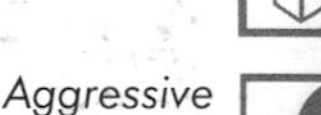

Savannah

Arboreal

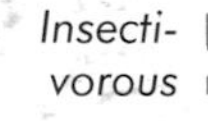

Aggressive

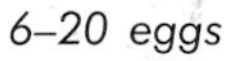

Insectivorous

6–20 eggs

This species is found in southern Europe and western Asia. The care of these popular lizards is similar to that of the preceding species. Males can be identified by their femoral pores, and they are more brightly colored than females, with deep blue throats when in breeding condition. It is usual to keep these lizards in groups containing one male and two or three females. Avoid housing males together, as this will lead to fighting. It is possible to accommodate these lizards outdoors in the summer in a sheltered and secure vivarium, designed with large stones which can serve as basking sites. This will also allow the lizards to synthesize vitamin D_3, but you will still need to provide extra calcium, in the form of a supplement or scrapings from a cuttlefish bone, especially at this time of the year, when females are likely to be laying eggs.

REPRODUCTION Cooling the lizards' quarters down to 50°F (10°C) in the late winter before raising the temperature again in the spring is essential. The female then lays her eggs approximately 6 weeks after mating. In an outdoor vivarium, it is especially important to watch for signs of laying activity, because the eggs may be lost once they are buried. Hatching takes about 2 months. The young green lizards, which measure about 3 in. (7.5 cm) at this stage, are predominantly brownish.

GREEN LIZARD
Green lizard is a general term used for members of the Lacerta genus.

GREEN IGUANA

Iguana iguana

These spectacular lizards from Central and South America are frequently sold as young hatchlings, which have been bred on special farms. Iguana meat is a popular dish, and the lizards are sometimes raised for this purpose by artificially hatching eggs. This allows some to be sold as pets and also provides stock to release back to the wild. There can be some variation in coloration in the case of green iguanas; while hatchlings generally are a much brighter shade of green than adults, some also have a distinct bluish hue to their bodies. These are highly prized, although their coloration may be temporary.

Before purchasing a young green iguana, it is important to realize that the relatively small hatchling, measuring perhaps 10 in. (25 cm), will soon grow into a much larger lizard, which can measure over 6 ft. (2 m) from its nose to the tip of its tail, and will clearly need much more spacious accommodation. In mild climates it will be possible to keep these iguanas in large outdoor vivaria, but otherwise, a large area of a room will be needed for them, and heating costs will multiply correspondingly.

60 in. (150 cm)

Tropical woodland

Arboreal

Aggressive

Vegetarian

20–40 eggs

Hatchling green iguanas are particularly susceptible to chilling and need to be kept at a relatively constant temperature, which must not fall below 77°F (25°C). A thermal gradient should be established across their quarters, by means of a spotlight which provides additional heat during the daytime, making sure they cannot burn themselves.

A large water bowl, the contents of which are changed daily, is necessary for bathing purposes and to assist in maintaining the humidity in the vivarium. Skeletal weaknesses, and even hind limb paralysis, are relatively common in the case of green iguanas. This can be due in some cases to a lack of suitable lighting in the iguanas' quarters, preventing them from metabolizing the calcium in their diet. An ultraviolet source above their favorite basking point is therefore essential, with a so-called "black light" being ideal.

A very varied diet of fruit and greens, such as bananas, grapes, and dandelions, should be provided. Young green iguanas will also eat

invertebrates such as crickets, which help to provide animal protein to support their growth. Such is the popularity of these lizards, however, that it is now possible to provide specially formulated diets that contain all the essential ingredients to keep them in good health. Some of these foods can be fed either in a dry state or with water added to them. Any surplus should be removed at the end of the day, in accordance with the feeding instructions on the packaging. Such foods can be mixed with fruits and vegetable matter. Separate formulations are available for young green iguanas, containing a slightly higher level of protein than the equivalent adult diet. Storing dry food presents no problems, even after the pack has been opened, but do not exceed the recommended "use by" date, as the vitamin level gradually decreases. It is also possible to buy canned diets for iguanas, which may be more palatable. These cans must be refrigerated after opening, although it is a good idea to allow the food to warm up for a few minutes before offering it to the iguanas. As with dry foods, both adult and juvenile formulations are available.

GREEN IGUANA
Green iguanas are suitable companions for large tropical turtles and tortoises.

The claws of green iguanas are sharp, and especially with tame individuals, they may need filing back, because they can cause a painful scratch and even damage furniture. Beware of cutting the nails, however, because they will bleed if they are cut too short.

Exercise is important to keep these lizards from becoming obese. Some owners like to exercise their iguana outdoors in sunny weather. Special harnesses for this purpose are available from pet stores specializing in reptiles. Even so, it is important to remember that your iguana could be easily scared by other animals, so never be tempted to tie the iguana's leash to a pole and leave it alone in unfamiliar surroundings.

REPRODUCTION Male green iguanas can be recognized by the larger crests running down their back. They become very territorial as they grow older. The female will lose her appetite for a week or longer before egg laying, but is likely to be increasingly thirsty. A relatively deep container of moist sand will be required for her nest, which may be 2 ft. (60 cm) deep. Incubation should take place at a temperature of around 86°F (30°C), and a relative humidity figure in excess of 80 percent. The young iguanas, averaging 10 in. (25 cm) long, will start hatching within 17 weeks. It will take them about 3 years to attain sexual maturity.

GREEN IGUANA
Tests have shown that it is possible for these lizards to distinguish between different colors.

DESERT IGUANA

Dipsosaurus dorsalis

15 in. (38 cm)

Desert

Terrestrial

Aggressive

Vegetarian

3–10 eggs

These attractively patterned iguanids originate from arid areas of south west USA and northern Mexico, where there is little natural cover available from potential predators. They can therefore run fast, even on a sandy surface, using their powerful hind limbs. They also burrow, retreating underground at night. In a vivarium, a suitable substitute for a natural burrow can be created with pipework set at an angle on the floor. Hot spots for basking and daily exposure to ultraviolet light are essential for these lizards. The bulk of their diet should consist of plant matter rather than fruit. Invertebrates such as crickets may also be eaten.

REPRODUCTION Relatively little is known about the reproductive behavior of these lizards. Males lack the enlarged scales located behind the anal region, which are present in females.

DESERT IGUANA
Desert iguanas may hibernate for part of the winter, burrowing into the substrate of the vivarium.

SIX-LINED RACERUNNER

Cnemidophorus sexlineatus

11 in. (27 cm)

Savannah

Terrestrial

Aggressive

Insectivorous

4–6 eggs

A member of the Teiidae family of lizards, the six-lined racerunner is so called because of the white lines running the length of its body. They come from eastern and central USA and, as their shape suggests, racerunners are fast-moving, active lizards which require a spacious vivarium. Part of the substrate should be moist, while a localized heat spot is also necessary, along with good ultraviolet lighting, since racerunners will often bask in sunshine. They are relatively shy by nature, and providing cork bark and other retreats should help them to settle in new surroundings.

This particular species will spend some of the winter hibernating, often retreating into an underground burrow. Lowering the temperature in their quarters by about 22°F (10°C) for a couple of months (provided that the lizards are in good condition) may stimulate breeding in the spring.

REPRODUCTION Male six-lined racerunners can be distinguished by their larger femoral pores and the presence of pores just in front of the vent. Mating occurs within a couple of weeks or so after the end of hibernation. The eggs are typically buried under suitable protection, such as cork bark, and should hatch after an interval of about 2 months.

SIX-LINED RACERUNNER
Even hatchling six-lined racerunners show the characteristic markings of this species.

CHINESE CROCODILE LIZARD

Shinisaurus crocodilurus

This highly unusual lizard was only discovered in 1930 and was virtually unknown to herpetologists outside China until the 1980s. It is so called partly because of the raised double row of scales running down the back to the tail, rather like those of a crocodile. In addition, unlike most other lizards, this particular species also appears to be semi-aquatic in its habits. Although the Chinese crocodile lizard is said to feed primarily on aquatic creatures, such as small fish and amphibians, it has proved possible to persuade them to eat more standard lizard fare, including crickets and earthworms. Avoid keeping these lizards at relatively high temperatures – a level just above room temperature suits them best. A localized heat source for basking is not recommended, although ultraviolet lighting should be included in their vivarium. The water here will need changing regularly, and it can help to include a separate container for this purpose, which simply needs to be lifted out, rinsed, and refilled.

REPRODUCTION Male crocodile lizards can be distinguished by their broader, more colorful heads, with an orange suffusion on the throat extending down the sides of the body. Females give birth to a small number of offspring, which are about 5 in. (12 cm) long. They should be transferred to separate quarters for rearing purposes, and a wide variety of foods provided to encourage them to eat properly.

12 in. (30 cm)

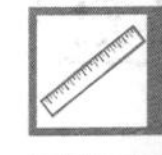

Semi-aquatic

Terrestrial

Aggressive

Carnivorous

2–12 live young

CHINESE CROCODILE LIZARD
The Chinese crocodile lizard is an unusual and interesting vivarium subject.

COLLARED LIZARD

Crotaphytus collaris

14 in.
(35 cm)

Desert

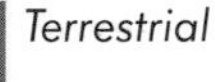
Terrestrial

Aggressive

Insectivorous

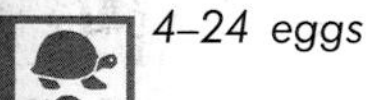
4–24 eggs

Males of this species rank among the most colorful of all the lizards found in the area from central USA to Mexico. Active by nature, collared lizards require a spacious vivarium if they are to thrive, with at least one basking point. A hot environment during the day, with the temperature falling at night, is recommended. Access to ultraviolet light is vital for their well-being.

Collared lizards are very agile and are capable both of running and jumping, which can make them difficult to catch if they escape from their quarters. In time, however, they can become quite tame, especially if you start with youngsters. Never keep collared lizards of widely different sizes together, because cannibalism is likely to occur.

Although they may attempt to bite when first handled, these lizards soon lose this tendency once they realize they will not be harmed. Their bite in most instances is not painful. Shaking the lizard gently will cause it to release its grip. Collared lizards are not adept climbers, although they naturally inhabit areas where rocks are found, hiding among them if they are frightened. Similar structures should be provided in their vivarium.

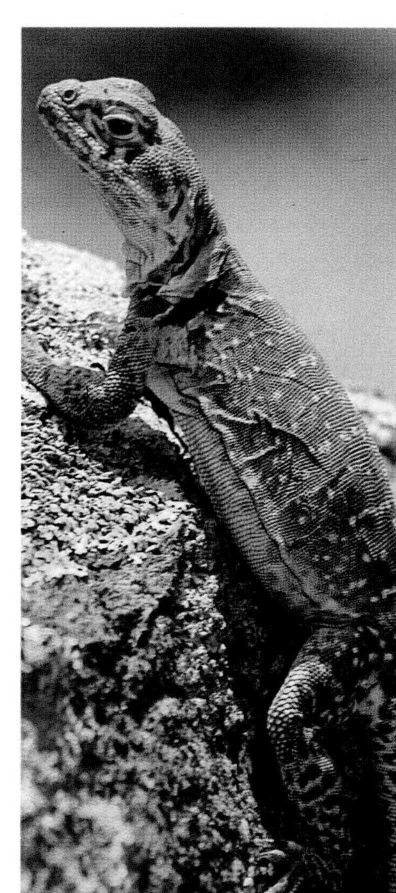

COLLARED LIZARD
Sexing collared lizards is straightforward.

REPRODUCTION While male collared lizards can be easily recognized by their brilliant greenish body coloration, with yellow markings, females of this species are predominantly brown. As the time for egg laying approaches, so red spots become apparent on the female's flanks, and she will swell with eggs. These are generally buried in a shallow scrape in the substrate, although they may be concealed under rocks on occasions. They are soft-shelled at this stage. Hatching is likely to take about 8 weeks. The young lizards can be reared on small crickets, augmented with other items such as spiders, which are a particular favorite.

WESTERN FENCE LIZARD

Sceloporus occidentalis

The scales in the case of this and other *Sceloporus* species are raised to a variable extent away from the body, particularly toward the tail, which is why they are also sometimes called spiny lizards. They require a similar set-up to collared lizards, with a slight decrease in temperature during winter being beneficial. They come from the USA and although they prefer relatively arid surroundings, water should always be available to them, in a dish on the floor of their quarters. Other related species, notably the green spiny lizard (*S. malachiticus*), which occur in forested areas, require more humid surroundings, achieved by regular spraying of the vivarium.

REPRODUCTION Females are relatively uniform in color, whereas males have blue underparts, which become obvious during display. Their eggs hatch within 7 weeks, with the young being just 2 in. (5 cm) long. Some other species of spiny lizard, typically those found at higher altitudes, produce live young, after a gestation period which can be between 7 and 10 months.

WESTERN FENCE LIZARD
The western fence lizard may lay its eggs in batches over several days.

9 in. (23 cm)

Savannah

Terrestrial

Aggressive

Insectivorous

6–13 eggs

CHUCKWALLA

Sauromalus obesus

One of the most striking of the lizards occurring in south west USA and Mexico, the chuckwalla is a member of the Iguanidae family. Although usually dark in color, chuckwallas can alter their coloration quite dramatically in response to their surroundings, with the tail often being lighter than the rest of the body. The chuckwalla's appearance is also influenced by its age, with young animals tending to be more colorful than adults. Their bodies are marked with a number of dark cross bands, which then start to disappear as the lizards grow older, being replaced by lighter areas that ultimately become reddish.

Chuckwallas live in rocky areas, and have evolved a unique means of protecting themselves here. If threatened, they will inflate their bodies, by breathing in more air than usual and using their tongues to prevent air from being lost from their lungs. This allows them to increase their volume by over 50 percent. These lizards are not difficult to maintain in vivarium surroundings, thriving on a herbivorous diet, supplemented with some fruit. Basking facilities and exposure to ultraviolet light in their quarters are essential, and the use of a vitamin and mineral supplement is also recommended.

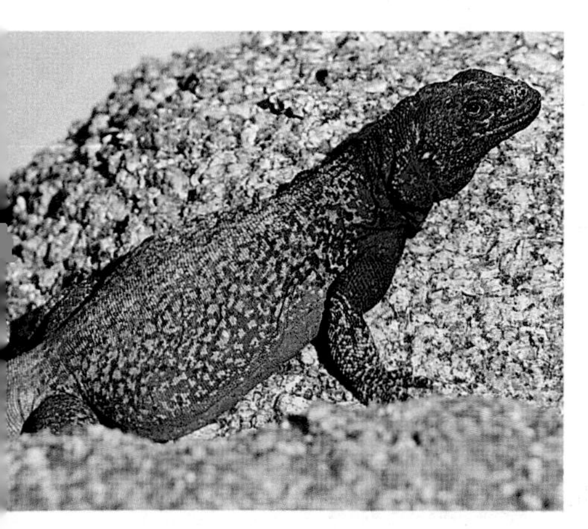

CHUCKWALLA
Chuckwallas are recognizable by their fat bodies, as reflected in their scientific name.

REPRODUCTION Adult females are smaller than mature males in general appearance, and they may have narrower heads. Young chuckwallas should emerge from their eggs after a period of 11 weeks or so, at an incubation temperature of 89.6°F (32°C). At this stage they will be about 2½ in. (6 cm) long.

18 in. (45 cm)

Desert

Terrestrial

Aggressive
Vegetarian

6–13 eggs

GREEN ANOLE

Anolis carolinensis

In spite of their name, the coloration of these small lizards can vary dramatically within minutes, depending on their surroundings. As a result, they are sometimes inaccurately called "American chameleons." as they originate from the southern USA. Their green coloration is actually an indication of a state of tranquility. If they are threatened or challenged by a rival, their color changes to a dark brown, almost verging on black in some cases.

Green anoles are an ideal choice for novice lizard keepers, being relatively inexpensive, with their small size making their care quite straightforward. The key element to keeping these anoles successfully is to provide a semi-moist environment; they will not thrive in dry surroundings. Regular spraying of the plants in their vivarium each evening will be adequate for this purpose. An ultraviolet light for basking is also essential. The temperature in the anoles' quarters should be lowered slightly at night.

Although green anoles will drink droplets of water from vegetation, they should also be provided with a drinker, such as one of the broad-based designs produced for pet birds. It is not uncommon for recently imported green anoles to be dehydrated, so you can place the lizard in a very shallow bath of water, to which a soluble vitamin and mineral tonic has been added. This can prove to be remarkably beneficial in a relatively short time. Change the water twice daily to make sure it is fresh.

9 in. (23 cm)

Tropical woodland

Arboreal

Aggressive

Insectivorous

2 eggs

GREEN ANOLE *Green anoles are mature by 9 months of age.*

REPRODUCTION Male green anoles have a larger dewlap in the vicinity of the throat, which they inflate as part of their display, or as a threat to other males, with which they may fight ferociously. If a female is ready to accept the advances of a male, she stays still, bending her head down. The male gains a hold by biting her neck so that mating can begin.

Egg laying takes place about 2 weeks later, with the female usually excavating a hole in the substrate for the eggs, digging it with her head and forelimbs. Incubation lasts for 8 to 12 weeks, with the young anoles being about 2 in. (5 cm) long at this stage.

BROWN ANOLE

Anolis sagrei

This highly adaptable species found in southern USA, the Caribbean and Central America is less commonly available than its green relative, but is equally easy to keep. The floor of the vivarium can be lined with bark chippings to create a more natural appearance than gravel, and plants in pots can be set here, with sphagnum moss used to disguise the containers if necessary. Reasonably sturdy, broad-leafed plants of appropriate size are suitable. Bromeliads are ideal, benefiting from the humidity. Various orchids may also be suitable and will not be harmed by the anoles, although if any of the plants flower, then the lizards may seek the nectar.

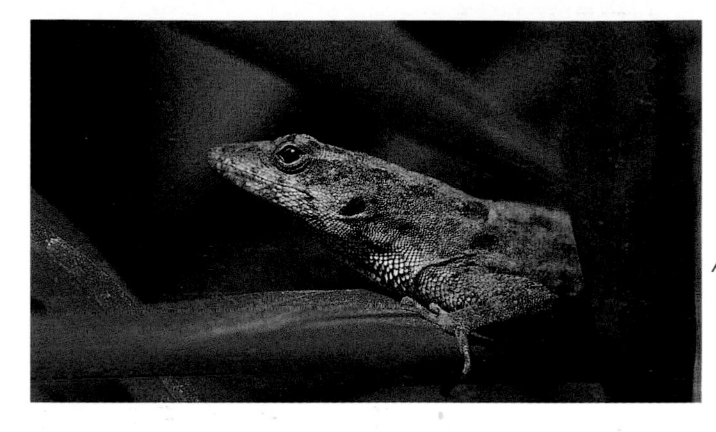

REPRODUCTION **Sexing and egg-laying details for this species are basically the same as for the green anole. Several clutches of eggs are likely to be laid in succession over the course of perhaps 4 months. The young are not difficult to rear on live foods of suitable size, with as much variety as possible being beneficial. Ultraviolet light is also essential for them.**

BROWN ANOLE
Attractive light brown patterning down the back is characteristic of the brown anole.

8 in. (20 cm)

Tropical woodland

Arboreal

Aggressive

Insecti-vorous

2 eggs

KNIGHT ANOLE

Anolis equestris

This species from Cuba is the largest member of the genus and can become quite tame, although if kept in a small vivarium, it is liable to injure its snout by rubbing itself on the front of the unit. Sometimes it may also be described as the Cuban anole. But unlike its smaller relatives, the knight anole is less active. Its diet also differs slightly, because it will eat pinkies as well as invertebrates, and may consume some vegetation. Plants included in vivaria provide camouflage, and substantial climbing branches should also be incorporated, in view of the size of these large creatures.

REPRODUCTION **The presence of two broad scales behind the vent serves to distinguish male knight anoles. They also have an impressive dewlap extending from the chin down to the shoulder region on the underside of the body. At a temperature of about 82°F (28°C) hatching of the knight anoles should occur about 10 weeks later, with the young anoles measuring up to 6 in. (15 cm) long.**

KNIGHT ANOLE
The coloration of the knight anole can vary from shades of green through to turquoise blue.

22 in. (55 cm)
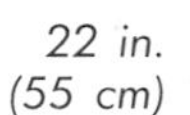

Tropical woodland

Arboreal

Aggressive

Insectivorous

1–2 eggs

NILE MONITOR

79 in. (2 m)

Savannah

Terrestrial

Aggressive

Carnivorous

10–60 eggs

Varanus niloticus

This species is found in Africa apart from the northwest. Young hatchlings are very colorful, but as they grow the attractive yellow markings tend to disappear, being broken by dull brown areas. It is important to know that the 8-in. (20-cm) long young will grow into large and potentially aggressive adults. These are not easy lizards to handle at this stage and can cause painful injuries. They are capable of inflicting severe bites, while their sharp claws can make deep scratches in bare skin. Last but not least, a blow from the tail of a fully grown Nile monitor can knock an adult off balance. Although they may appear sluggish for long periods, these large lizards can move with surprising speed. Monitors in general are lizards for the specialist, being more suited to zoological collections because of the large amount of space they need as they grow.

REPRODUCTION Most Nile monitors are kept alone, so captive breeding is rare. Males tend to be larger and will inflate their gular pouch in the vicinity of the throat as part of their display. Incubation typically lasts for 20–30 weeks at a temperature of about 86°F (30°C).

NILE MONITOR
The Nile monitor is one of the most colorful, but least friendly, species.

BOSC'S MONITOR

69 in. (1.75 m)

Savannah

Terrestrial

Aggressive

Carnivorous

10–50 eggs

Varanus exanthematicus

Small examples of this species, which is also known as the savannah monitor, are regularly available and need similar care to that required by the Nile monitor. Larger invertebrates will be eaten by young Bosc's monitors, in addition to pinkies and other animal foods. It is possible to purchase special canned diets for them from specialist reptile suppliers. As they grow, the size of their food should be increased accordingly. These monitors are greedy feeders, and excess food will ultimately result in obesity, shortening their lifespan. The tail is used to store fat, being rounded in a well-fed Bosc's monitor. Fresh eggs are a particular favorite, and these monitors can swallow them whole without breaking the shell. Originating from west and central Africa, Bosc's monitor is not a keen swimmer, but it still needs a large bowl of water for bathing.

REPRODUCTION Pairs need to be introduced to each other cautiously, as fighting may occur. Females lay their eggs in termitaria in the wild, which offer protection from predators. In vivarium surroundings, raised areas of firm sand should provide an acceptable substitute. Hatching takes about 170–180 days, with the young monitors being 3 in. (8 cm) long when they emerge from their eggs.

BOSC'S MONITOR
Design quarters for these lizards so they can be cleaned easily, with minimum disturbance.

BEARDED DRAGON

Pogona vitticeps

Large numbers of these attractive lizards from Australia are now being bred in many countries around the world. Their coloration may vary quite widely, with some individuals showing few dark markings through to others which are nearly black. The name "bearded" refers to the spiny projections around the chin and throat area, which also extend down the sides of the body.

A basking area is important in a vivarium for these large lizards, as is regular exposure to ultraviolet light. A drop in temperature at night is also recommended. Only relatively tough plants should be included because they may be eaten.

Quite social by nature, bearded dragons will communicate with each other by means of their body language. There is normally a dominant male in each group that enjoys priority during feeding, and also in other situations such as selection of basking sites. Social status begins early in life, and is based on the size of the lizards, with bigger hatchlings being dominant.

Although these agamids conceal the full extent of their beards, when they are at rest, they can be fanned out as a threat to intimidate a possible opponent. Turning one of the front legs is a gesture of appeasement, preventing any further display of aggression at this stage. When fighting does occur, one of the combatants will seek to pin down its opponent by biting the back of the neck. At this point, the weaker individual backs down by lying on the ground, and no further conflict occurs. Although these encounters may seem serious, they rarely result in injury.

REPRODUCTION Male bearded dragons have pores extending from in front of the anus over the femoral area at the top of each leg. They display by bobbing their heads in front of the female who, if she is receptive, will lower her body. A deep container should be provided for the eggs, which may be laid in batches over a month. Nearly 30 may be produced, and they should be transferred to an incubator. Hatching occurs about 3 months later, with the young being about 3 in. (7.5 cm) long. Successful rearing requires regular ultraviolet exposure, together with an adequate intake of calcium for bone growth.

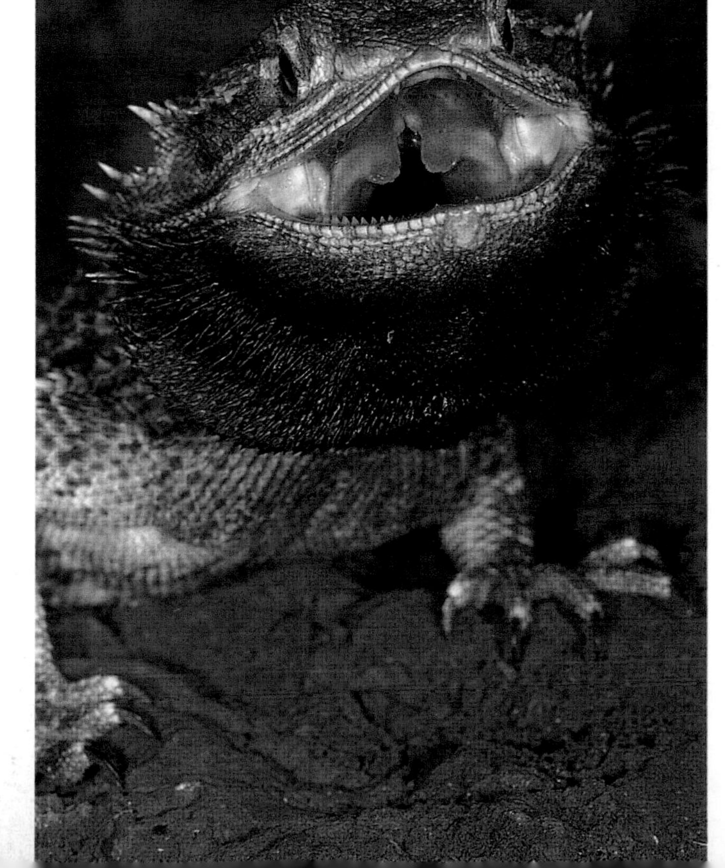

BEARDED DRAGON
Bearded dragons are lizards with personality and can become very tame, especially if obtained young.

20 in. (50 cm)

Desert

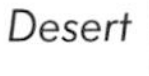

Terrestrial

Agree in groups

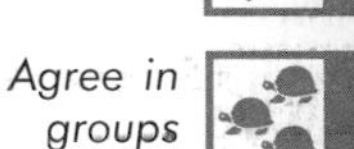

Insecti-vorous

15–30 eggs

COMMON AGAMA

Agama agama

The common agama comes from the savannah areas of North Africa. Like many lizards, agamas will change their color according to their mood, and the temperature of their surroundings. Males become especially colorful when displaying, with the heads turning a vivid shade of orange-red. The pores near the vent also serve to distinguish them from females. These lizards are highly territorial, and males should be kept apart because, once they settle in their surroundings, the dominant individual will bully its companions. Their enclosure must incorporate retreats, because of the rather nervous nature of these lizards.

16 in. (40 cm)

Savannah

Terrestrial

Aggressive

Insectivorous

10–20 eggs

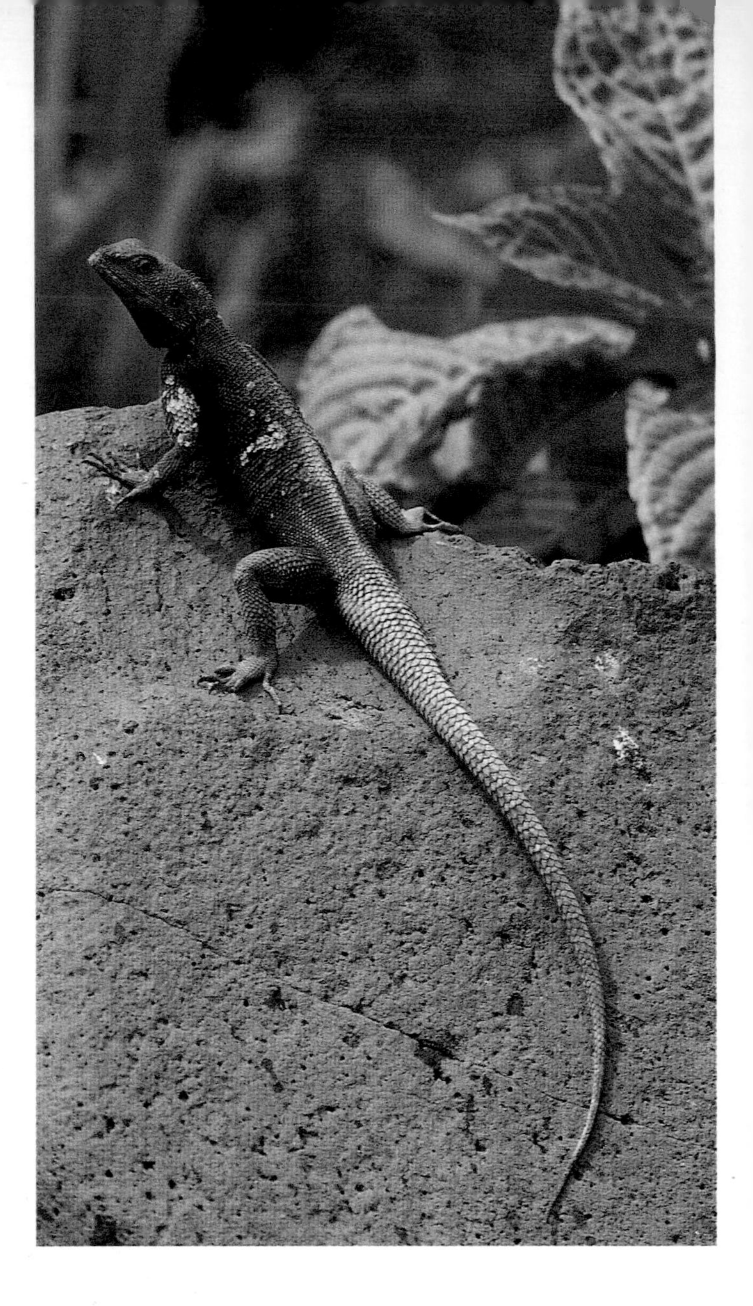

COMMON AGAMA
These agamas are active lizards, requiring spacious surroundings.

REPRODUCTION Pronounced head bobbing is an integral part of the agama's display. Should the female raise her back and stand upright, this is a sign that she is receptive to the male's advances. He will clasp her neck in his jaws during mating; this is quite normal behavior at this stage. The agamas' quarters must be equipped with a suitable container in which the female lizard can dig a hole and lay her eggs. At a temperature of 86°F (30°C), the incubation process can vary from about 6 weeks to nearly 2 months.

FIVE-LINED SKINK

Eumeces fasciatus

9 in. (22 cm) 
Temperate woodland 
Terrestrial 
Aggressive 
Insectivorous 
15 eggs 
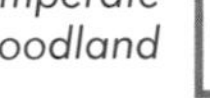

These lizards come from eastern USA and are also known as blue-tailed skinks. The five-lined skink is sometimes confused with similar species from Africa. The bright blue color of the tail is especially vivid in youngsters and serves as a defense mechanism, drawing predators away from their head. The tail is quite easily shed, allowing the skink to escape danger. It will regrow but may be shorter than before. Take care when handling these agile skinks to prevent this occurrence. Moist areas and retreats such as tree bark should be provided in a vivarium. Males must be kept apart from each other because they are aggressive by nature and can injure themselves if they fight. They can be distinguished by their more colorful heads.

REPRODUCTION Female five-lined skinks will conceal their eggs in a suitably damp locality in their vivarium, with incubation lasting for approximately 6 weeks. During this time, the female lizard is likely to remain close to her eggs. In the case of some other *Eumeces* species, the female will actually incubate the eggs by curling around them, which is unique in this group of reptiles.

FIVE-LINED SKINK
Cooling the living quarters of five-lined skinks in late winter may stimulate breeding in the spring.

RED-TAILED FLAT ROCK LIZARD

Platysaurus guttatus

8in. (20 cm)  
Savannah 
Arboreal 
Aggressive 
Insecti-vorous 
2–4 eggs 

The flattened shape of these lizards allows them to squeeze into rock crevices and escape possible danger, and their vivarium needs to be designed to allow them to climb on rocks. It is vital to arrange the rockwork so there is no risk of collapse, which could be fatal to the lizards, or which could allow them to escape if it smashes the front of the vivarium. Light artificial rockwork sold for use in aquaria is a safer option than real stone. These lizards come from southern Africa and will avidly search for crickets and similar invertebrates which will move over the rock faces. A special vitamin and mineral mix should be sprinkled over the crickets to improve their nutritional value. Ultraviolet lighting and a heat spotlight are essential for rock lizards, which are active during the day.

REPRODUCTION Sexing is very straightforward in this case, since only males have the characteristic red tail. Females are predominantly brown, with a dull yellow stripe running down their back. They are also smaller in size. Their eggs are hidden among rocks. The young should then emerge about 6 weeks later.

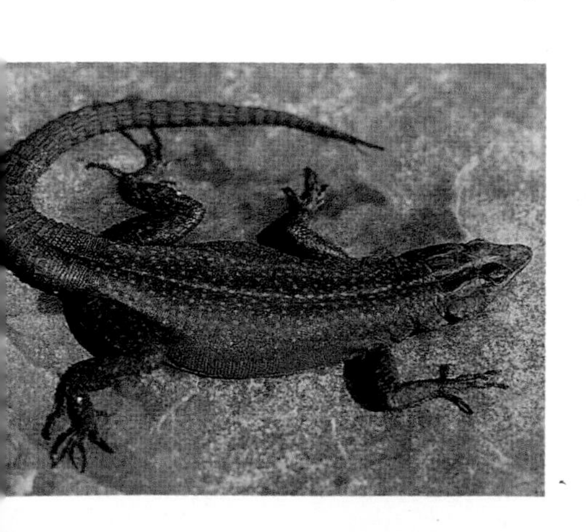

RED-TAILED FLAT ROCK LIZARD
Males of this species have especially attractive coloration.

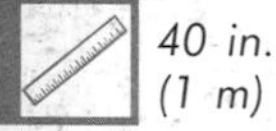
40 in.
(1 m)

Desert

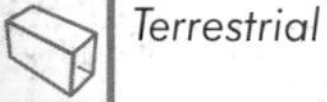
Terrestrial

Aggressive

Carnivorous

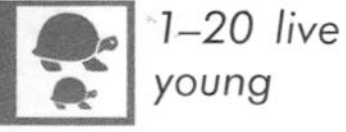
1–20 live
young

ROUGH-SCALED SAND BOA

Eryx conicus

The rough-scaled sand boa is the most attractively marked member of its genus, with some individuals having a distinctly dark zigzag pattern of markings running down their back. Their underparts are pale. The raised scales of this snake give it its name. It inhabits arid areas of India, Pakistan, and Sri Lanka, and will frequently burrow during the day. Its vivarium must therefore have a deep substrate, such as horticultural sand, allowing the snake to conceal itself. Securely positioned rockwork will provide further retreats at ground level. No other vivarium structure is necessary, since these snakes show no inclination to climb. They should, however, have a water bowl. Feeding at dusk normally draws the snake quickly from its hiding place. Mice are suitable for adult individuals.

REPRODUCTION **Reducing the humidity and lowering the temperature by several degrees appear to act as conditioning factors with rough-scaled sand boas. Mating should take place soon after standard conditions are restored, with the young being born about 4 months later.**

ROSY BOA

Lichanura trivirgata

40 in. (1 m) 

Savannah 

Terrestrial 

Aggressive 

Carnivorous 

3–10 live young 

In spite of the name, there is noticeable variation in the coloration of these snakes. They are found in an area from western North America to Mexico. Those of the desert subspecies, *L.t. gracia*, vary from pink to tan, while the coastal rosy boa (*L.t. roseofusca*) is of a deeper rosy shade, and is sometimes recognized as a species in its own right. These boas spend most of their time on the ground, although they may climb occasionally, so secure branches should be included. They are usually more active after dark. Young mice provide a good diet for rosy boas, which are a relatively small species. During the late winter, the temperature in their quarters should be lowered to stimulate breeding during the following spring.

REPRODUCTION Rosy boas often breed quite readily, with females giving birth after a period of at least 15 weeks. The young will grow relatively quickly on a diet of pinkies at first. They measure about 12 in. (30 cm) on hatching, and will be mature within 2 years, having doubled their body length.

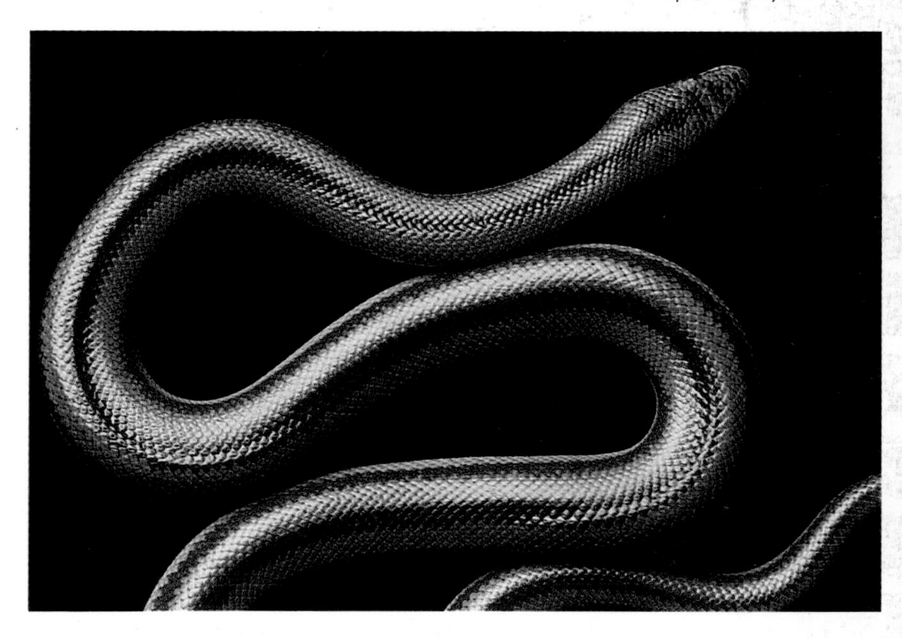

ROSY BOA
The coloration of individual rosy boas can vary quite widely.

ROUGH-SCALED SAND BOA
Sand boas lose their appetite when in breeding condition.

120 in. (3 m)

Savannah

Terrestrial

Aggressive

Carnivorous

6–22 live young

HAITIAN BOA

Epicrates striatus

This species is also known by other names, including Fischer's boa and the Antillean boa. Eight different subspecies have been identified on various Caribbean islands. So there can be a significant variation in appearance between snakes described as Haitian boas.

These snakes are essentially nocturnal, emerging to hunt at dusk. The Haitian boa is less demanding in its choice of prey than some other members of the genus. In addition to small mammals and birds, it will catch various lizards, although established specimens will eat typical snake fare such as mice and day-old chicks without problems. It can prove harder to persuade wild-caught individuals to eat such foods. Haitian boas tend to spend most of their time on the ground, but will climb on occasions so secure branches should be provided.

HAITIAN BOA
Epicrates boas are sometimes called slender boas, because of their body shape.

REPRODUCTION Lowering the temperature slightly and increasing the humidity in late winter will stimulate breeding in spring. The young, measuring up to 20 in. (50 cm) long, are sometimes reluctant to eat foods such as pinkies, although some will eat fish, lizards, and tree frogs.

80 in. (2 m)

Savannah

Terrestrial

Aggressive

Carnivorous

1–23 live young

RAINBOW BOA

Epicrates cenchria

This boa is found over a vast area of Central and South America, so a range of subspecies has evolved, with some such as the Brazilian rainbow boa (*E.c. cenchria*) being more colorful than others. This is usually reflected in the price of captive-bred stock, with such snakes being correspondingly more expensive. The dark markings in this case, forming circles or interconnecting loops down the back, are set against a reddish-orange background color. In other instances, however, the contrast is far less pronounced. It is important to maintain the characteristics of these different subspecies when captive-breeding, so purchase stock from a knowledgeable supplier.

Rainbow boas are highly adaptable snakes. They should, however, be provided with a secure water container which is deep enough to allow them to immerse themselves. Mice and rats of suitable size can form the basis of their diet.

REPRODUCTION Reducing the temperature slightly for several weeks in late winter should serve to stimulate breeding activity. The gestation period lasts for 6–7 months, with the young snakes averaging about 8 in. (20 cm) when they hatch.

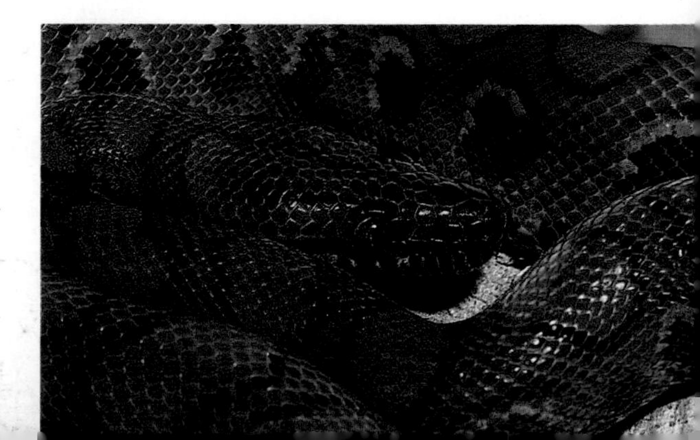

RAINBOW BOA
Rainbow boas are usually attractively marked, whatever their coloration.

COMMON BOA

Boa constrictor

120 in. (3 m)

Savannah

Arboreal

Aggressive

Carnivorous

8–40 live young

There is quite a wide variation in the coloration and patterning of these boas, depending on their area of distribution. Found in Central and South America, those from the southern part of their range, in Argentina and Paraguay, are considerably darker overall than those from elsewhere. It should be noted that Central American boas can be very aggressive.

In spite of their potential size, common boas are widely kept. They are not difficult snakes to keep, although they require a relatively large vivarium, taking into account their eventual large size. They will climb, and so some secure branches should be provided in their quarters.

A large water container is essential, both for bathing and drinking. Ideally it should be connected to a drain, as it will need to be cleaned and refilled daily. The floor of their quarters should incorporate a suitable retreat and also areas of softer materials such as sphagnum moss. Common boas will also bask, particularly under a spotlight, so take care that they cannot burn themselves here.

The temperature in their vivarium can be allowed to fall slightly at night, although young boas are susceptible to chilling and need to be kept at a relatively constant temperature. Mammals tend to be their favored prey, with large boas consuming rabbits as well as rodents. Birds may also be featured in their diet. Like many snakes, these boas tend to be at their most active from dusk onward, typically hunting at night.

REPRODUCTION Breeding tends not to follow a set cycle, but cooling the snakes' quarters in winter for between 6 and 8 weeks is likely to stimulate mating, assuming they are in good condition. These boas attain sexual maturity at approximately 3 years old. The male shows his interest by pursuing the female around their quarters, and when he gets close to her, he uses his anal spurs to caress her body as a prelude to mating. The young are born up to 35 weeks later. At this stage, they may measure 20 in. (50 cm) in length. They should be reared in a separate vivarium, feeding at first on pinkies within 5 weeks of birth.

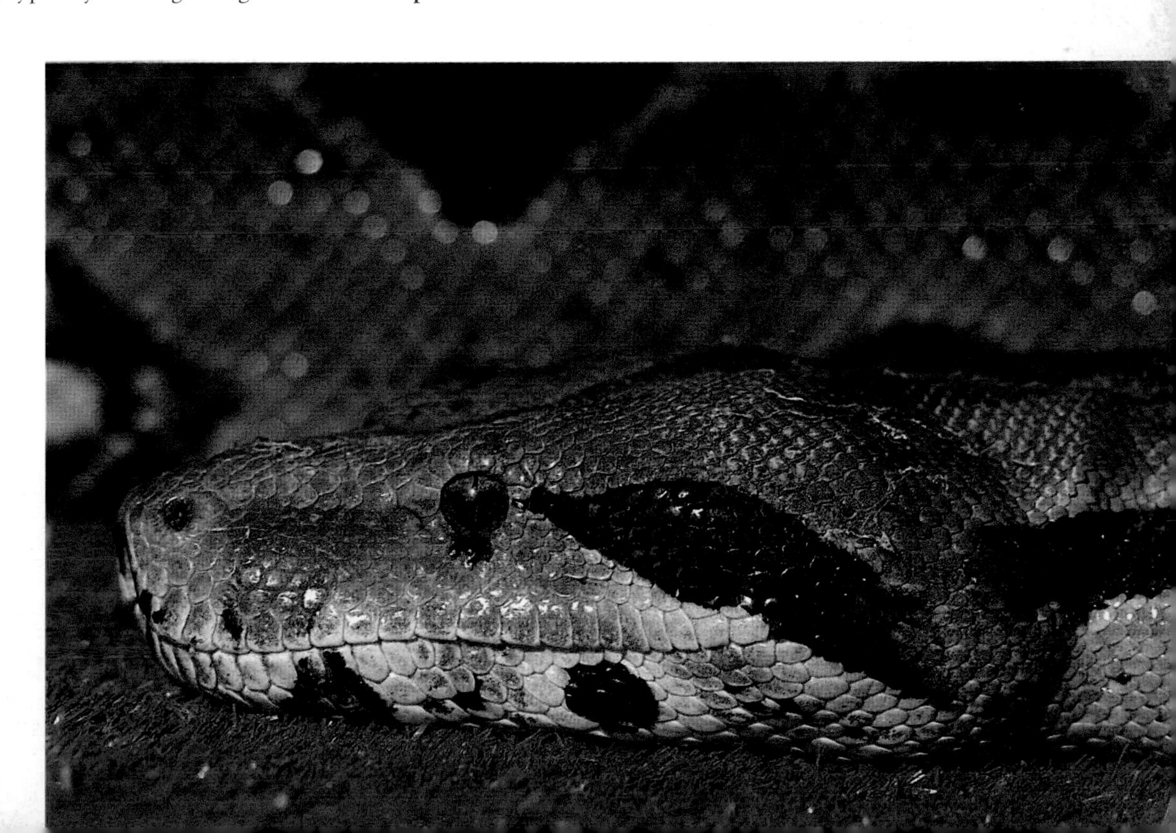

COMMON BOA
Although they can be difficult when larger, these boas have reasonably friendly natures.

60 in. (1.5 m)

Tropical woodland

Arboreal

Aggressive

Carnivorous

1–15 live young

EMERALD TREE BOA

Corallus caninus

The beautiful coloration of these snakes helps to conceal their presence in the tree tops where they live in northern South America. They rarely descend to the ground, hunting birds and small mammals among the branches. This specialist lifestyle means that they need a relatively tall vivarium, with a variety of securely positioned branches for climbing. A feeding shelf is also useful in some cases, although these snakes can also be fed by using safe forceps. They are usually quite aggressive and generally difficult to maintain by amateurs.

Emerald tree boas will rest hanging off a horizontal branch and lap at water that collects in their coils. In a vivarium, a hook-on water dish, as sold for birds, can be mounted securely on the side. Netting staples partially driven into the side will act as holders for the hooks. Relatively robust climbing plants such as *Philodendron* can also be included in the vivarium, to give more cover.

EMERALD TREE BOA *The white markings distinguish the emerald tree boa from the green tree python.*

REPRODUCTION Spraying their vivarium more frequently to raise the humidity, which should be around 50–80 percent, and then lowering the temperature slightly should stimulate breeding. Young will be born about 6 months later, and should be moved to other quarters for rearing. They may be yellowish or reddish in color and mature at 2 years old.

48 in. (1.2 m)

Savannah

Arboreal

Aggressive

Carnivorous

2–7 eggs

BALL PYTHON

Python regius

Also known as the royal python, this species from western and central Africa is frequently available. Its accommodation needs are reasonably easy, although a water container large enough to allow them to bathe is important. Ball pythons will climb and need securely mounted branches in their quarters. These are not always easy snakes to establish, and it is important to see them feeding if possible. They can survive for months without eating. While fasting is normal to some extent, if it is protracted, it will leave the snake vulnerable to infection.

Ball pythons that are reluctant to feed on mice may be persuaded to take dead gerbils instead. By rubbing the mice with a gerbil to transfer the scent, you may persuade the snake to eat dead mice, which are easier and cheaper to obtain. Do not leave the snake too long before seeking veterinary assistance if it persists in refusing food, in case there is an underlying health problem.

REPRODUCTION These pythons are not frequently bred. Mating is likely during the spring in the northern hemisphere. The female will lay her eggs about 4 months later, and curls around them, protecting them from predators. Here she will remain until they hatch, after at least 2 months. The young pythons are then about 15 in. (38 cm) long.

BALL PYTHON *The ball python curls into a ball if threatened. This snake is about to shed its skin. Note the milky appearance of its eye.*

GREEN TREE PYTHON

Chondropython viridis

This species is found in New Guinea and northern Australia. Most green tree pythons are emerald green, with small markings down their back, but some individuals that have lost the yellow pigment from their skin are sky blue. A tall vivarium, with plenty of branches for climbing, is essential. The humidity should be high, maintained by daily spraying, with a water container set off the ground. Commonly, the snakes prefer to consume water caught in their coils. Food, such as day-old chicks and rodents, should be offered with blunt-nosed forceps, and even a wild snake should soon accept this method of feeding.

REPRODUCTION Breeding can be difficult. Females left to incubate their own eggs must have seclusion, and a nest box in which to lay their eggs on a bed of moist sphagnum moss. The young pythons should hatch from about 11 weeks, measuring about 10 in. (25 cm). They can be reluctant to eat and may have to be force-fed at first. Their color can vary from bright yellow through to a reddish shade.

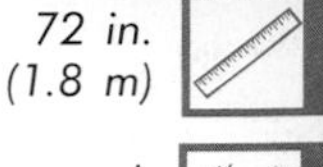
72 in. (1.8 m)

Savannah

Arboreal

Aggressive

Carni-vorous

12 eggs

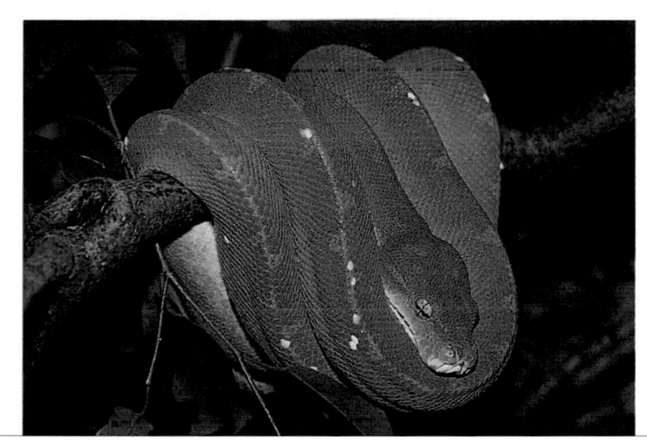

GREEN TREE PYTHON
These pythons will rest by looping themselves around a horizontal branch.

CHILDREN'S PYTHON

Liasis childreni

This particular species from northern Australia has become more commonly available in recent years, thanks to captive breeding. It is a relatively small python and can be kept without great difficulty. Climbing and bathing facilities should be provided in its vivarium, with moss and sand used as a lining. Children's pythons will often hide when they are not hunting, so a suitable retreat will also be required in their quarters. These pythons will normally eat mice readily, often preferring to feed at night, especially in unfamiliar surroundings. They have a very keen sense of smell which guides them to their food, even when it is totally dark.

CHILDREN'S PYTHON
Children's python is a species growing in popularity.

REPRODUCTION Reducing the temperature in the vivarium for a couple of months in the winter should trigger breeding, with the pythons being kept together for mating purposes at this stage. Eggs are laid just over 5 months later, with incubation itself continuing for about 9 weeks. These pythons are small when they hatch, often measuring little over 8 in. (20 cm) at this stage.

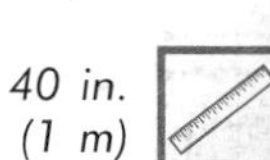
40 in. (1 m)

Savannah

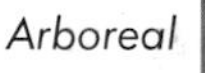

Arboreal

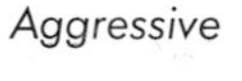
Aggressive

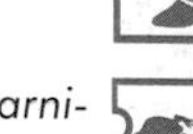

Carni-vorous

3–14 eggs

RETICULATED PYTHON

Python reticulatus

The huge potential size of these snakes, coupled with their often rather unpleasant nature, means that they are not a good choice for most snake keepers, especially novices. Reticulated pythons are found in south east Asia. They grow very fast, being second only to the anaconda in size. They therefore require spacious surroundings, which are likely to be available only in a zoological collection. Space heating is then normally used to warm their quarters, with localized hot spots for basking. A pool for bathing is also necessary. Feeding presents few problems, as reticulated pythons will eat a wide variety of animal food, ranging from rats and rabbits to chickens in the case of the larger individuals. Always take care when feeding or cleaning out the quarters of snakes of this size, as they can be aggressive toward people.

360 in. (9 m)

Tropical woodland

Arboreal

Aggressive

Carnivorous

30–100 eggs

REPRODUCTION This is not generally achieved in captivity, largely because of the demands of keeping these snakes. They are mature when they reach a length of 11½ ft (3.5 m). Cooling their quarters stimulates mating behavior. Egg laying takes place between 3 and 5 months later. Young reticulated pythons measure approximately 30 in. (75 cm) long when they hatch after a further 2 months.

RETICULATED PYTHON
The reticulated python often attains a fearsome size.

EGG-EATING SNAKE

Dasypeltis scabra

24 in. (60 cm)

Savannah

Arboreal

Aggressive

Carnivorous

2–15 eggs

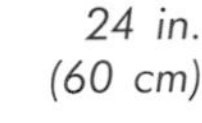

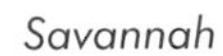

This most common of the six related species of egg-eating snake comes from Central and Southern Africa. Although not especially striking in terms of coloration or patterning, these snakes have fascinating feeding habits. As their name suggests, they feed on eggs, finding birds' nests by a combination of smell and taste via their tongues. Egg-eating snakes have flexible jaws which allow adults to consume whole hens' eggs without difficulty. Once the egg is in the snake's throat, sharp projections from the vertebrae above split the shell, allowing the contents to run into its stomach. The shell is later voided from the mouth.

Egg-eating snakes are found in a wide range of habitats, but they prefer relatively dry conditions. Sand, part of which is kept moist, can be used to line the floor of the vivarium, and rocks and branches for climbing should also be included Ultraviolet lighting is beneficial. A container of water is essential.

EGG-EATING SNAKE
These snakes grate their scales to make a hissing sound if they are threatened.

REPRODUCTION Mating typically takes place in July, with egg laying occurring a month later. The eggs will be buried in the substrate by the female, and the young should hatch roughly 8 weeks later. They will measure just 8 in. (20 cm) at this stage, and a source of suitably small eggs, from prolific aviary birds such as the zebra finch (*Poephilia guttata*) or Bengalese (society) finch (*Lonchura domestica*) will be needed. As they grow larger, quail's eggs, which are more readily available, can be provided.

100 in. (2.5 m)

Savannah

Terrestrial

Aggressive

Carnivorous

3–15 eggs

BULL SNAKE

Pituophis melanoleucus sayi

This snake is the largest subspecies of the widely distributed gopher or pine snake (*P. melanoleucus*), and is also the biggest snake found in North America. Its coloration is highly variable, which is probably a reflection in part of its wide distribution. Wild bull snakes can be rather aggressive and difficult to handle in view of their size, whereas young captive-bred hatchlings usually prove to be fairly docile.

Bull snakes are relatively easy to maintain, though a spacious vivarium is needed because of their large size. A fairly arid enclosure, equipped with a suitable retreat and some branches, will suit them well. Feeding is very straightforward, with dead mice being taken readily.

GOPHER SNAKE
There are about 15 different subspecies, ranging from Canada to Central America.

REPRODUCTION Cooling the snakes' quarters to create an impression of winter is a vital conditioning stimulant. Mating takes place in the spring, and the female will lay roughly 1 to 2 months later. Incubation then typically lasts a further 2 months, with the hatchlings being relatively large and with good appetites, being reared easily on pinkies. Young bull snakes grow rapidly and should be mature by 2 years old. Care and breeding requirements for other subspecies are similar.

80 in. (2 m)

Savannah

Arboreal

Aggressive

Carnivorous

5–20 eggs

AMERICAN RACER

Coluber constrictor

These active, fast-moving snakes found in an area from south west Canada to north east Mexico benefit from being kept in a spacious vivarium where they can also climb. They bask readily, and both localized heat spots and ultraviolet lighting should be provided. Their coloration can vary through their wide range, with some individuals being melanistic, although they generally tend to be shades of green.

American racers are nervous and dislike being handled. They are not difficult to feed on a diet based on small mammals and larger insects, although in the latter case, additional vitamins and minerals will be valuable. In areas where the climate is favorable, a secure outdoor vivarium provides the best way to keep these snakes, with additional heating if necessary. This will allow them to sun themselves and lessens the risk of snout lesions caused by the snake rubbing itself on the front of a vivarium.

REPRODUCTION Male American racers can often be recognized by their longer tails, compared with those of females. Breeding in captivity is not common. The young are usually grayish, with darker spots and bands when they hatch, and measure 8 in. (20 cm) or so in length. Invertebrates may prove to be a satisfactory first food.

AMERICAN RACER
These snakes use their speed to catch lizards.

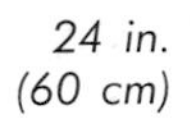

WESTERN HOG-NOSED SNAKE

Heterodon nasicus

24 in. (60 cm)

Savannah

Terrestrial

Aggressive

Carnivorous

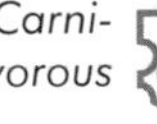

9–21 eggs

The distinctive flattened nasal area of these snakes has given rise to their popular name. If scared, they can flatten the skin of their neck region, to create an impression of a cobra, hissing alarmingly at the same time. Should this not deter a potential predator, the hog-nosed snake then becomes inert in the hope that this will cause it to be left alone, as many animals instinctively respond to movement, and will not scavenge for prey.

Found in an area from southern Canada to eastern Mexico, in the wild, these snakes tend to prey on toads, but they are adaptable and can usually be persuaded to eat dead mice. They must have bathing facilities and retreats in their quarters.

REPRODUCTION It is important to reduce the temperature at which these snakes are kept, taking it down to about 59°F (15°C) in the late winter, as a means of conditioning them to breed in the spring. The hatching rate is normally very good, with the incubation period itself lasting about 7 weeks and the hatchlings measuring about 6 in. (15 cm) at first. Pinkies make a satisfactory rearing food. The snakes will be mature within about 2 years.

WESTERN HOG-NOSED SNAKE
These snakes can live into their teens.

20 in. (50 cm)

Savannah

Arboreal

Social

Insectivorous

5–12 eggs

SMOOTH GREEN SNAKE

Opheodrys vernalis

Both species of green snake from North America are easily managed, and make highly attractive vivarium subjects. They have the advantage of being insectivorous – unlike most snakes. Suitable foods can be obtained from specialist suppliers. Crickets, wax moth larvae, and mealworms are all suitable. Nevertheless, these snakes are at greater risk of suffering from nutritional deficiencies than their carnivorous cousins. Regular use of a vitamin and mineral supplement, preferably a balancer sprinkled over the invertebrates, is recommended, as is ultraviolet lighting in the vivarium to assist synthesis of vitamin D_3. Valuable variety can be added to the diet by offering other invertebrates, such as spiders, obtained by trawling suitable undergrowth with a cheesecloth net, away from areas where insecticides may have been used. Daily feeding is recommended.

SMOOTH GREEN SNAKE
Smooth green snakes produce elongated eggs.

REPRODUCTION Similar management to that recommended previously for other snakes from this part of the world is necessary to stimulate breeding. The eggs must be taken out of the vivarium for incubation, because otherwise they will be attacked and destroyed by the crickets. Hatchling crickets will serve as a rearing food in due course.

48 in. (1.2 m)

Semi-aquatic

Terrestrial

Can be kept in groups

Carnivorous

5–28 live young

WESTERN RIBBON SNAKE

Thamnophis proximus

Ribbon snakes are closely related to garter snakes, but they can be distinguished by their thinner body shape. There are six distinctive subspecies of the western ribbon snake over its extensive range from central USA to Costa Rica. The stripes down the back can vary from light brown through yellow to red, and individuals vary widely. Ribbon snakes live close to areas of water, particularly where there is vegetation to conceal their presence. They hunt both amphibians and fish.

These snakes require an area of water in their vivarium, with moss and suitable retreats also present. It is vital that part of the substrate is dry, because otherwise they are likely to succumb to ulcers on the underside of their bodies. A relatively large enclosure is needed for these active creatures. Handling ribbon snakes is reasonably straightforward, although they can bite in spite of their small size. Feeding should be as for the related garter snake (see page 79).

REPRODUCTION Keeping these snakes in groups appears to encourage breeding. They will hibernate together, with the temperature being lower in their quarters during the winter, and then mate almost as soon as they emerge in the spring. The young will be born after 4 months, with the female often giving birth in damp moss.

RIBBON SNAKE
Ribbon snakes, especially hatchlings, have slim bodies, and their vivarium must be kept secure.

COMMON GARTER SNAKE

Thamnophis sirtalis

This is the most widely distributed snake in North America, where it is usually found near water. As might be expected, there can be considerable variation in appearance between garter snakes from different parts of the continent. They are often recommended as a species suitable for novice snake keepers, but although they are not difficult to maintain, garter snakes do have specific requirements. At first, they may produce a very unpleasant secretion from their cloacal glands when they are handled, so it is always a good idea to keep the rear end of the snake well away from your clothing.

Fish are normally prominent in the diet of these snakes, but unfortunately, commercially available sources such as whitebait contain a harmful enzyme called thiaminase. This serves to de-activate the vitamin B_1 (thiamin), with the result that the garter snake will start to develop signs of a deficiency. These symptoms begin with loss of coordination, followed by progressive paralysis, which is ultimately fatal. Cooking the fish lightly and letting it cool before offering it to the snakes will de-activate the enzyme, and a source of vitamin B, such as brewer's yeast, can also be added to the fish. The fish should not be fileted, in the same way that whole prey must be provided to other snakes, in order to meet their requirements for minerals, notably calcium. Some garter snakes can also be persuaded to sample earthworms, which can be obtained from specialist live food suppliers.

A recent innovation in the feeding of both garter snakes and ribbon snakes has been the development of special complete diets for them. These contain all the necessary ingredients to keep them in good health, and provide a much better option than relying on fish as the main constituent of their diet.

REPRODUCTION Breeding requirements are similar to ribbon snakes. Young garter snakes measure about 7 in. (18 cm) and should be reared away from adult snakes. They should be ready to breed in about 2 years.

COMMON GARTER SNAKE
These snakes are quite easy to handle.

52 in. (1.3 m)

Semi-aquatic

Terrestrial

Can be kept in groups

Carnivorous

7–85 live young

52 in.
(1.3 m)

Savannah

Terrestrial

May eat smaller companions

Carnivorous

5–14 eggs

MILK SNAKE

Lampropeltis triangulum

The unusual name of these snakes originates from the myth that they used to feed on cow's milk, possibly because they were seen in pastureland. Their appearance varies, even between related individuals. Found in an area from eastern USA to northern South America, those from the southern part of its distribution are the most colorful. As a result the Central American subspecies, such as the Pueblan milk snake (*L.t. campbelli*) originating from Mexico, are most commonly kept and bred. This particular subspecies also has relatively large hatchlings, which has increased its popularity, as they can be reared on a diet of pinkies.

Selective breeding of color variants, as distinct from recognizable subspecies, has given rise to the tangerine variant of the Honduran milk snake (*L.t. hondurensis*) among others, in which the white bands have been replaced by tangerine coloration. This can be so dark that it becomes indistinguishable from the snake's red areas in some cases. The care of these snakes does not differ significantly from that of other *Lampropeltis* species.

MILK SNAKE
In spite of its prominent red coloration, which mimics the deadly coral snake, the milk snake is not a poisonous species. This is the Sinaloan form.

REPRODUCTION Although Central American forms tend to need a slightly higher temperature, they are still equally responsive to cooling in the winter to stimulate breeding. Mating takes place from spring through to early summer, with the eggs then taking approximately 2 months to hatch.

72 in.
(1.8 m)

Savannah

Terrestrial

May eat smaller companions

Carnivorous

3–21 eggs

COMMON KINGSNAKE

Lampropeltis getulus

The basic yellow and black of these snakes is seen in various combinations through their range. They are found in southern USA and Mexico. In the case of the eastern subspecies (*L.g. getulus*), the blackish-brown areas predominate, with thin streaks of yellow. Bigger yellow bands are evident in California kingsnakes (*L.g. californiae*), while a mottled pattern is characteristic of the Florida subspecies (*L.g. floridana*).

Kingsnakes can appear quite threatening if cornered, rearing up, rattling their tail like a rattlesnake, and then lunging quickly to inflict a passing bite. This would give them the opportunity to escape in the wild, retreating to nearby cover.

Like other *Lampropeltis* species, kingsnakes will not climb, so a fairly low vivarium is suitable. Their quarters should correspond with the habitat of the subspecies being kept. A water bowl is essential in all cases.

REPRODUCTION It is vital to pair these snakes carefully, or the unique characteristics of individual types will be lost. The male starts courtship by trying to bite his mate, until he can gain a grip to climb onto her back. This process may take about 4 hours. Young kingsnakes hatch after about 2 months. It is safest to rear them separately because of cannibalism.

COMMON KINGSNAKE
The eastern kingsnake, showing the characteristic markings of this subspecies.

MEXICAN KINGSNAKE

Lampropeltis mexicana

MEXICAN KINGSNAKE
The Mexican kingsnake is one of the tricolored species.

40 in. (1 m)

Savannah

Terrestrial

May eat smaller companions

Carnivorous

3–15 eggs

Found in a mountainous area of Mexico, these Mexican kingsnakes show the same type of variability in appearance as other species of kingsnake. They all have a distinct three-colored pattern, with gray as the ground color, which has given rise to their alternative name of gray-banded kingsnake, although this name is also applied to another species, *L. alterna.* There can be further confusion over these kingsnakes, because the different subspecies are also sometimes referred to under separate names. These include the San Luis Potosi kingsnake (*L.m. mexicana*), Thayer's kingsnake (*L.m. thayeri*), and the Durango mountain kingsnake (*L.m. geeri*). They require a reasonably arid environment, complete with hiding places, as they can be rather shy, especially at first. As a group, these snakes are primarily nocturnal, becoming active at dusk and hiding away during the day.

REPRODUCTION Reducing the vivarium temperature for about 2 months during the winter is essential to ensure successful breeding of this species during the following spring. It should be maintained at 59°F (15°C). Clutch size appears to depend to some extent on the subspecies concerned – the Durango mountain kingsnake tends to lay 6 eggs per clutch, whereas females of the San Luis Potosi subspecies may produce double this figure. Hatching takes roughly 8–12 weeks.

BLACK RAT SNAKE

Elaphe obsoleta

With colorful snakes tending to be favored, the normal form of this rat snake has become less popular. Instead, the amelanistic variant, which is now well established, is more likely to be seen in collections. The typical form is often entirely black, with an attractive glossy sheen.

Many rat snakes show a variation in color between adult and young, and up to nine different subspecies are recognized. Sometimes young retain the patterning that distinguished them as hatchlings – a grayish shade, with darker markings down their bodies.

Found in an area from eastern North America to northeastern Mexico, the most distinctive form is the Everglades rat snake (*E.o. rossalleni*), found in southern Florida. It is an orange shade, with slightly darker stripes. This subspecies is relatively arboreal in its habits. In central and northern parts of Florida, extending into Georgia and the Carolinas, the yellow rat snake (*E.o. quadrivittata*) can be found. These snakes have been nicknamed chicken snakes because of their appetite for eggs, although they also catch rodents regularly. Hatchlings often hunt amphibians, which can make them tricky to rear.

72 in. (1.8 m)

Savannah

Arboreal

May eat smaller companions

Carnivorous

6–14 eggs

The different subspecies of the common rat snake overlap, and where they meet they may hybridize. There is also a distinct green phase of the yellow rat snake in the north of its range. The oak snake, also called the gray rat snake (*E.o. spiloides*), is similar to the so-called Gulf Hammock rat snake (*E.o. williamsi*), to the extent that not all taxonomists recognize the division between them.

The gray rat snake is often attractively marked, with dark blotches contrasting with a background color that may vary from brown through shades of gray almost to white. There is little difference between the appearance of young and adults. Their local name of "oak snake" comes from the fact that they are often found in oak trees. They climb readily – which should be reflected in the design of the vivarium.

A similar but less distinctive patterning characterizes the Texas rat snake (*E.o. lindheimeri*), which occurs in eastern and central parts of Texas, as well as through Louisiana and western Arkansas. Some individuals have a reddish hue, with their scales being edged accordingly.

Farther south, ranging from the south of Texas down into Mexico, is Baird's rat snake (*E.o. bairdii*), another distinctive member of the species. It tends to be smaller than the Texas rat snake, as well as being less sharply patterned. Baird's rat snake is sometimes considered to be a separate species. It is relatively rare.

As might be expected with such an adaptable species, it is relatively easy to keep in fairly dry surroundings. It will not thrive under humid or damp conditions. Although these snakes will feed on a variety of foods, they can be maintained very satisfactorily on a diet consisting mainly of rodents.

REPRODUCTION Cooling during late winter is important for breeding success. Mating takes place in spring, with the eggs being laid approximately a month later. Incubation lasts about 2 months, with the hatchlings averaging 12 in. (30 cm) long. Females sometimes lay a second clutch of eggs in late summer.

BLACK RAT SNAKE

Rat snakes are very variable in terms of their coloration and markings.

CORN SNAKE

48 in. (1.2 m)

Savannah

Arboreal

May eat smaller companions

Carnivorous

8–26 eggs

Elaphe guttata

The relative ease of care and the readiness of these snakes to reproduce under captive conditions has given them widespread popularity. They are found in an area from south eastern North America to eastern Mexico. Regional variants occur, with corn snakes originating from Okeetee in South Carolina being especially vivid and brightly colored. While the most common species has reddish saddle-like patterning edged with black down its back, the Great Plains subspecies (*E.g. emoryi*) has these red areas replaced by brown. In contrast, the rosy corn snake (*E.g. rosacea*) from southern Florida tends to lack the black border around the red marking, although in many cases there is variation in this respect.

Various strains of mutation corn snakes are now well established as the result of captive breeding. One of the most striking is the so-called "Blood red," where the entire body has a red suffusion. There are also colorful amelanistic corn snakes, where the dark pigment is missing, resulting in orange, red, and cream areas. "Snow corns" result when these colors are also missing, creating snakes that are pure white.

Corn snakes are secretive by nature and should be housed in a vivarium with adequate hiding places. Bathing facilities should also be offered but their quarters should be dry.

CORN SNAKE
Corn snakes are also sometimes called red rat snakes.

REPRODUCTION **Breeding requirements are similar to those of the common kingsnake (see page 80).**

RUSSIAN RAT SNAKE

60 in. (1.5 m)

Savannah

Arboreal

May eat smaller companions

Carnivorous

9–14 eggs

Elaphe schrencki

Rat snakes are not only confined to the Americas; they are also found in Central Asia, although these species are less widely kept. One reason is that they tend to be less colorful, and there can be difficulties in establishing wild-caught individuals. They are often imported with a relatively heavy burden of parasites, which may not be directly life-threatening, but will usually depress the snake's appetite, so that it may refuse to feed. Treatment at the outset is therefore recommended (see page 34). Once these rat snakes are feeding, there should be little to worry about, as they are not otherwise difficult to keep, and can be bred quite easily. They normally kill their prey by constriction, but will eat dead rodents, especially mice, as well as the occasional day-old chick.

RUSSIAN RAT SNAKE
An attractive bluish sheen is visible on these snakes in sunlight. Yellow markings become apparent by 18 months old.

REPRODUCTION **Female Russian rat snakes may lay their eggs quite late in the year, with development of the embryos actually starting while the eggs are in the female's body. Hatching takes about 6 weeks. The young measure about 12 in. (30 cm) long and are far less colorful than adults at this stage. They can be reared on pinkies.**

44 in.
(1.1 m)

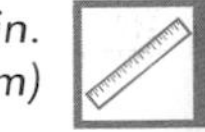

Savannah

Terrestrial

Aggressive

Carnivorous

2–15 eggs

BROWN HOUSE SNAKE
The brown house snake is an attractive species whose care is straightforward.

BROWN HOUSE SNAKE

Lamphrophis fuliginosus

There is some confusion over the scientific name of this snake, which used to be known as *Boaedon fuliginosus.* The brown house snake comes from southern Africa and is a relatively easy species to keep, requiring a dry vivarium and suitable hiding places, although it is not especially shy, as it is often found in or near homes. Brown coloration predominates in this snake, but there may be quite wide variations in its depth. The underparts are a pale shade of cream, with pale stripes also running along the sides of the body, up to the head and passing through the eyes.

Feeding should not normally present any problems, with rodents providing the basis of the diet. In the wild, these snakes will often hunt lizards and they are able to take relatively large prey.

REPRODUCTION It is not uncommon for female brown house snakes to lay two or more clutches during the summer, following a winter period when the temperature in their quarters has been reduced. The eggs will hatch after an interval of about 10 weeks, with clutch size depending to some extent on the size of the snake. Smaller females produce fewer eggs. Young brown house snakes can be 10 in. (25 cm) long, and are quite easy to rear on pinkies.

80 in. (2 m)

Savannah

Arboreal

Aggressive

Carnivorous

4–12 eggs

INDIGO SNAKE

Drymarchon corais

INDIGO SNAKE
Indigo snakes are protected in many countries including the USA, and so traffic in this species is difficult

The coloration and patterning of this snake vary through its wide range, to the extent that eight different forms are recognized by some taxonomists. The Florida subspecies (*D.c. couperi*) is probably most widely kept. It is a sleek but not brightly colored snake, being a glossy black, sometimes with a bluish tone, with deep salmon or white markings on the sides of the head extending to the chin. In contrast, young indigo snakes have a banded appearance. If alarmed, these snakes will vibrate the tip of their tails, in a similar way to rattlesnakes. They will also prey on these deadly snakes, although in a vivarium they will feed on foods such as dead rodents.

REPRODUCTION Male indigo snakes are likely to become very aggressive when they are in breeding condition. Matings should be supervised, with the female being transferred elsewhere afterward. Unfortunately, although indigo snakes lay quite readily, the level of fertility may be low, in spite of the fact that females can retain viable sperm in their bodies. Repeated matings may help to increase the level of fertility. The typical incubation period for the eggs is about 3 months, with the young measuring up to 24 in. (60 cm) on hatching.

PRAIRIE KINGSNAKE

Lampropeltis calligister

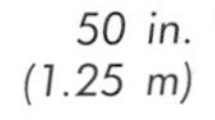

50 in. (1.25 m)

Savannah

Terrestrial

May eat smaller companions

Carnivorous

6–17 eggs

Three distinct subspecies of the prairie kingsnake occur through its range in south eastern North America, including an apparently isolated population in Florida, which was discovered approximately 200 miles (320 km) south of the nearest known population and described for the first time in 1987. The prairie kingsnake itself (*L.c. calligister*) is predominantly grayish-brown, with blotches running down the back of its body, and small dark areas intervening along the sides. The background color is usually yellowish-brown, but it can be significantly darker in some cases. There is also a well-established amelanistic variant, which has dull orange markings against a pale salmon background color. The third race, known as the mole snake (*L.c. rhombomaculata*), occurs further east, and tends to lose its patterning as it matures, becoming dark brown much like the color of a mole. It is far less commonly kept and bred, probably because of its dull coloration, although feeding may also prove more difficult. Some reports suggest that this species prefers to hunt other reptiles, including snakes, rather than small mammals.

REPRODUCTION Breeding is similar to other kingsnakes. The young average 10 in. (25 cm) after an incubation of approximately 2 months, and are normally spotted in appearance. They need to be reared individually, as with related species.

How long will snakes live? Can I age an adult snake with any degree of reliability? *Most snakes are likely to have a life expectancy of a decade or more in captivity, although this will often be shorter in the wild where they face many dangers. Unfortunately, it will be impossible to determine the age of an adult snake with any degree of certainty. It is often better to start with hatchlings therefore, since you can be certain of their age.*

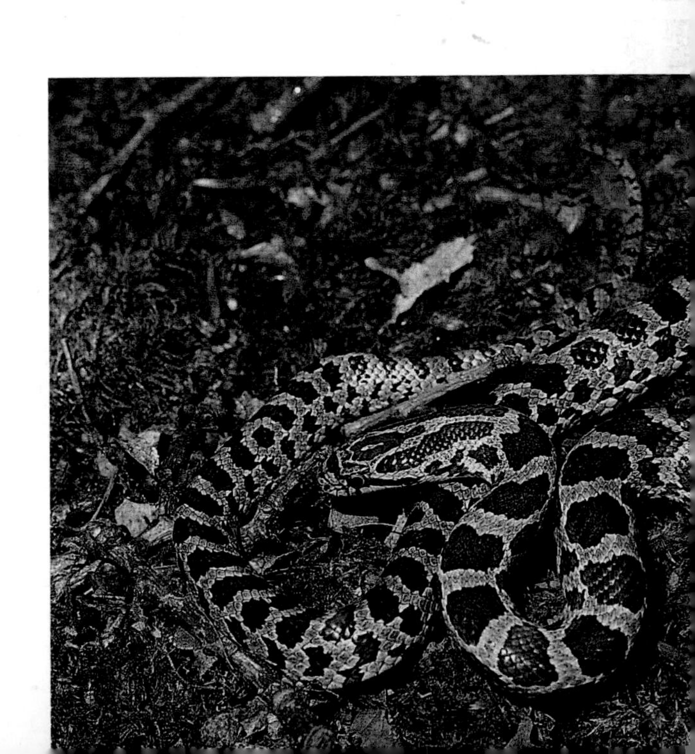

PRAIRIE KINGSNAKE
This species has proved to be one of the easiest kingsnakes to keep and breed in vivarium surroundings.

28 in. (70 cm)

Savannah

Terrestrial

Agree in groups

Vegetarian

4–15 eggs

LEOPARD TORTOISE

Geochelone pardalis

These tortoises show considerable variation in their markings, with some individuals having an attractive balance between the light cream and darker blackish areas on their shells. Their coloration tends to become duller as they grow older.

A dry enclosure is vital to their well-being, although a dish must also be provided for drinking water. Cool, wet surroundings are likely to cause respiratory infections which can prove fatal, especially in younger animals.

Leopard tortoises are found in central and southern Africa. They have large appetites, and although they generally prefer herbage, they also eat fruit. Provide a wide variety of vegetation, particularly shoots of plants such as dandelions (*Taraxacum officinale*), although they will also eat the flowers. During winter in temperate climates, you may need to grow food for them, fresh beansprouts being suitable. Avoid giving too much cabbage and other brassicas, because these contain a compound which can affect the thyroid gland, and may interfere with the tortoise's metabolism.

LEOPARD TORTOISE

Leopard tortoises should only be allowed outside on warm sunny days.

REPRODUCTION Males, distinguished by their flat plastrons (undersides), will chase females in typical tortoise fashion. Breeding is likely once the females have grown to about 15 in. (38 cm) long. Incubation time can be very variable. It is likely to be at least 3 months, and can take 9 months according to some reports. The ideal temperature is 86°F (30°C).

8 in. (20 cm)

Savannah

Terrestrial

Agree in groups

Vegetarian

3–4 eggs

BELL'S HINGEBACK TORTOISE

Kinixys belliana

This is one of three species of hingeback tortoise from central and southern Africa, and it is the most adaptable. It is found in a wide range of habitats, whereas the other hingebacks live mainly in areas of tropical forests. The characteristic hinge on the upper side of the shell allows them to draw their hind legs totally into their shell for additional protection. The appearance of Bell's hingeback can differ from one individual to another, some having shells with no patterning, whereas others may display considerable variegation. The shell is relatively low, giving these tortoises an elongated appearance. This may give them some protection against predators, since they can hide under rocks. They prefer to have retreats of this type in their vivarium. Hingebacks generally dislike bright sunlight, but an ultraviolet light is important for their health.

In the wild, these tortoises may estivate for periods, especially in dry weather. If a newly acquired specimen is reluctant to feed, place it in a shallow pan of tepid water to which a vitamin solution has been added. These tortoises all drink relatively large volumes of water. Herbage and fruit will be taken, under normal circumstances, with tomatoes often being a favored food.

REPRODUCTION The tails of males are significantly longer than those of females. The eggs are buried and will have to be moved carefully to an incubator. At a temperature of approximately 86°F (30°C), they should hatch in about 18 weeks. The young tortoises measure about 1½ in. (4 cm) in length.

BELL'S HINGEBACK TORTOISE

Hingebacks have a flattened top to their carapace.

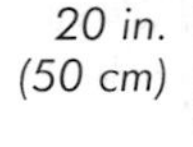
20 in.
(50 cm)

Tropical
woodland

Terrestrial

Agree in
groups

Vegetarian

6–15 eggs

RED-LEGGED TORTOISE

Chelonoidis carbonaria

These relatively large tortoises from tropical South America have orangish markings in the center of each scutum on their carapace, as well as along the edge of the shell in some cases. Reddish areas are also evident on the front legs and the head, although the depth of this coloration varies noticeably between individual tortoises, with some being significantly duller than others. A very similar species is the yellow-footed tortoise (*C. denticulata*), which is found only east of the Andes. It has markings that are more yellowish, but this division is not always easy to make. Lack of pigmentation on the underside of the red-footed tortoise's shell can be a helpful additional means of distinguishing them.

A warm, reasonably humid environment is required by these tortoises, with a water bowl large enough for them to rest in. There is little point in including plants, however, because these are likely to be eaten or simply destroyed if they are within the tortoise's reach. Fruit tends to be preferred over greens, with plums being a particular favorite. They may sometimes take animal food, such as canned dog food.

RED-LEGGED TORTOISE
Spacious surroundings are needed for these tortoises.

REPRODUCTION **Breeding can occur at any stage of the year. The female red-footed tortoise will bury her eggs in a hole slightly over 8 in. (20 cm) deep. Incubation in this instance can last 6 months, the young red-footed tortoises being nearly 2 in. (5 cm) long when they hatch.**

MEDITERRANEAN SPUR-THIGHED TORTOISE

Testudo graeca

12 in. (30 cm)

Savannah

Terrestrial

Agree in groups

Vegetarian

4–13 eggs

This species originates from countries around the Mediterranean. It used to be one of the best-known tortoises, with thousands being sold as pets each year. Sadly, many died as the result of exposure to unsuitable climatic conditions. They either succumbed directly to being chilled, or failed to eat adequately to survive hibernation. Since then, a much better insight into their needs has been acquired, and captive breeding is now commonplace.

The care of hatchlings is different from that of adult tortoises, particularly in terms of their dietary needs. Mediterranean spur-thighed tortoises will eat a wide variety of plant matter, although they tend not to eat grass. Dandelions and clover are particular favorites, with both leaves and flowers being consumed readily. Green vegetables such as alfalfa can be offered as a substitute when other foods may be in short supply. There is now a general tendency among breeders especially not to provide their tortoises with any form of meat, although it is not uncommon for these reptiles to catch and eat snails and earthworms if the opportunity presents itself outdoors. Excess amounts of cat food have been linked with shell deformities in hatchlings, because of the high level of protein, along with other harmful effects.

MEDITERRANEAN SPUR-THIGHED TORTOISE
The Mediterranean spur-thighed tortoise can be easily identified by an enlarged conical scale.

Tortoises of all ages, but particularly youngsters, must have adequate calcium carbonate in their diet, to prevent them from succumbing to osteodystrophy. Typical signs of an impending problem are a relatively soft shell, coupled with scutes on the carapace growing vertically into a raised pyramid shape, instead of being rounded. There are now specialist calcium supplements available for reptiles, although grated pieces of cuttlefish bone (as sold for pet birds) are equally good. This can be mixed in with chopped green stuff if necessary, but tortoises may consume it by itself quite readily. Effective absorption of calcium depends on vitamin D_3, and so exposure to ultraviolet light, to allow the manufacture of this vitamin, is essential.

Food should be withheld in the autumn before hibernation, because otherwise it may rot in the tortoise's gut. Tortoises must not be exposed to freezing conditions, and prefer a temperature of about 40°F (5°C). If the tortoise has not been eating properly, it is inadvisable to allow it to hibernate. Hatchlings can hibernate without problems, providing they are in good condition.

The care of Hermann's tortoise (*T. hermanni*), a Mediterranean species, is similar.

REPRODUCTION Female tortoises do not need to mate regularly to produce fertile eggs, as they can store sperm. They bury their eggs, and if they are living outside, close supervision is needed to spot this activity, which lasts at least 2 hours. The eggs are likely to hatch between 2 and 4 months later, depending on the incubation temperature.

HORSFIELD'S TORTOISE

Testudo horsfieldi

9 in.
(23 cm)

Savannah

Terrestrial

Agree in
groups

Vegetarian

4–5 eggs

This species is found in relatively inhospitable terrain in Central Asia, where summers are hot and dry, and winters are long and cold. These tortoises eat the spring vegetation and then retreat to underground burrows as the temperature rises. They hibernate through winter, so they are only active for about 4 months of the year. Horsfield's tortoises are far less readily available than the better-known Mediterranean species.

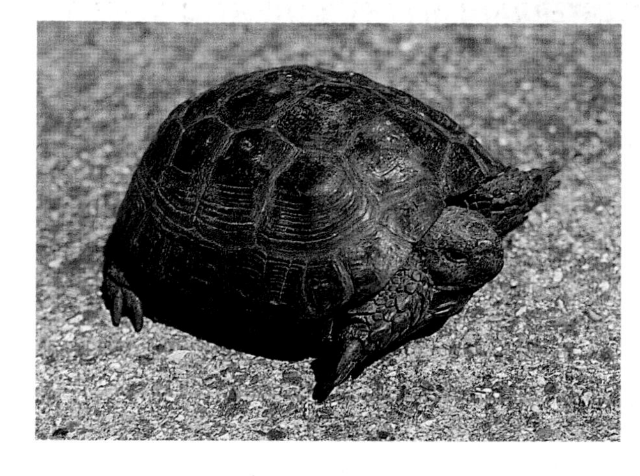

HORSFIELD'S TORTOISE
The flattish, rounded shape of the shell assists these tortoises as they burrow underground.

These tortoises should only be allowed out on dry sunny days, when they will eat a variety of greens, including grass. In summer, they bury themselves for progressively longer periods during the day. Horsfield's tortoises are able to climb wire mesh with confidence, so they must be kept in a fairly secure enclosure. After hibernation they will drink readily in the spring.

REPRODUCTION Little is documented about their breeding habits, although females lay during their period of summer activity. It takes between 80 and 100 days for the eggs to hatch.

SPINY TURTLE

Heosemys spinosa

9 in.
(23 cm)

Semi-
aquatic

Terrestrial

Agree in
groups

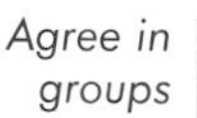

Omnivorous

Unknown

These chelonians are sometimes described as cogwheel turtles, because of the characteristic projections on the sides of their bodies. These tend to be most prominent in smaller turtles, and it has been suggested that they serve as protection from attacks by snakes, which could otherwise swallow them without difficulty. Spiny turtles come from Southeast Asia and require reasonably humid surroundings, and although they may not spend long periods in water, they will immerse themselves, and can swim quite powerfully. An area of sphagnum moss in the dry area of their vivarium will provide them with somewhere to conceal themselves on land. They tend to become more active as dusk approaches. They can be aggressive and may bite.

REPRODUCTION These turtles can be sexed by the longer tails of males, which also have the anogenital opening located farther away from the shell than females. Captive breeding is rare at present.

Unlike other turtles, the spiny turtle appears to feed primarily on land, taking fruit, especially banana, quite readily, often to the exclusion of other foods. To provide a more balanced diet, turtle foodsticks should be stuck into the pieces of banana. The turtles will soon eat these on their own, having acquired a taste for this unfamiliar foodstuff. Ultraviolet lighting also appears beneficial to them.

SPINY TURTLE
The projections around the sides of the turtle's body have sharp points.

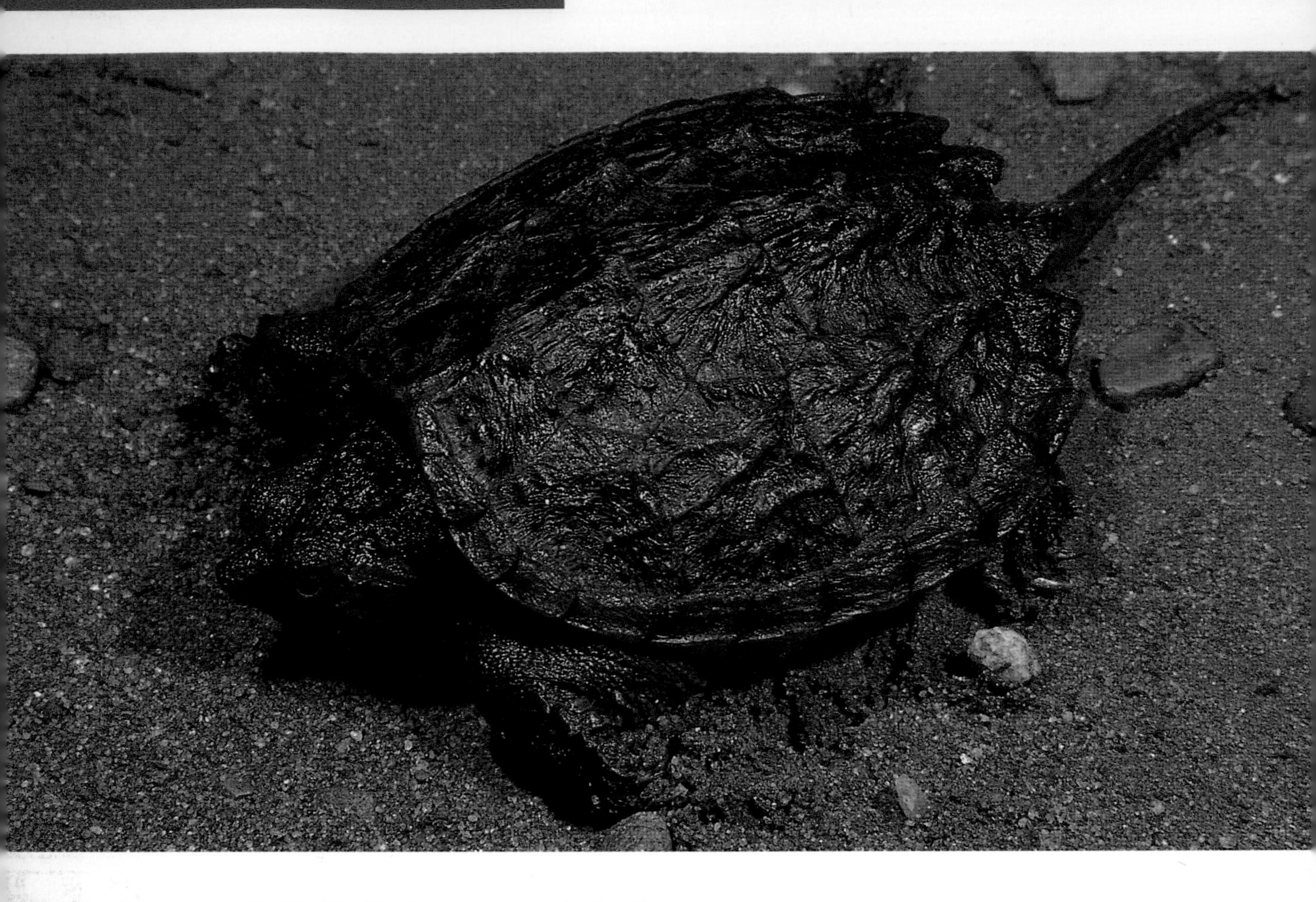

COMMON SNAPPING TURTLE *Snapping turtles can grow to a large size.*

 *19 in. (48 cm)*

 *Semi-aquatic*

 *Terrestrial*

 *Aggressive*

 *Carnivorous*

 *20–40 eggs*

COMMON SNAPPING TURTLE

Chelydra serpentina

This species is found in an area from southern Canada and eastern USA to Mexico. Hatchlings are occasionally available, but it is important to remember that they grow rapidly to a large size, soon outgrowing their accommodation. They spend a lot of time underwater, rarely emerging to bask, and are quite sluggish, although they actively resent being handled. Large individuals need to be lifted carefully by the front and rear of their shell, so there is less risk of being bitten. They may also release a foul-smelling substance from glands near the base of the tail, as another deterrent to being picked up.

Bigger snappers may take rodents as part of their diet. Hatchlings can be given turtle foodsticks.

REPRODUCTION The anogenital opening on the underside of the tail is closer to the shell in female snapping turtles. The underside of the shell also tends to be slightly broader than in males, whereas the top of their shell is relatively flat, to allow the male to maintain a grip around the edges during mating. Nesting usually peaks in the summer months, with incubation lasting about 3 months. Young hatchlings are surprisingly aggressive. They have rougher shells than adults.

COMMON MAP TURTLE

Graptemys geographica

The markings of hatchling map turtles are especially prominent, and the pattern of lines on their shells, resembling the contours on a map, explains their name. Another feature is the raised projections running down the center of their vertebral scuta, which are more prominent in some cases than others. They have given rise to their alternative popular name of "sawbacks," since these swellings can resemble the teeth of a saw. They can be found in an area from southern Canada south to Tennessee and Kansas.

Map turtles are moderately aquatic, but they will emerge from the water to bask in sunlight, or under an ultraviolet source. Rearing presents few problems, provided that the young turtles receive a properly formulated complete diet. Avoid those turtle foods which consist only of dried insects.

REPRODUCTION Sexing is straightforward in this species on the basis of size, males being significantly smaller than females. They may only grow to half the size of females, and have smaller heads. Hatchlings are all of similar size, however, measuring about 1½ in. (4 cm).

COMMON MAP TURTLE
The patterning of these turtles fades as they grow older.

9 in. (23 cm)
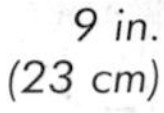

Semi-aquatic

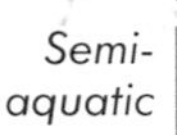

Terrestrial
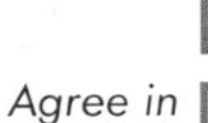

Agree in groups

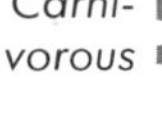

Carnivorous
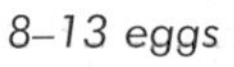

8–13 eggs

PAINTED TURTLE

Chrysemys picta

There are four distinct subspecies of the painted turtle, named after their distributions from southern Canada and central USA, although there is some overlap between them, with the eastern (*C.p. picta*) and midland (*C.p. marginata*) sharing part of their range. The eastern subspecies can be distinguished by its clear yellow plastron, whereas the midland has a dark area running down its center. The southern painted turtle (*C.p. dorsalis*) is the most distinctive, with an orange stripe running right down the center of its carapace. The western subspecies (*C.p. belli*) is potentially the largest and has yellow streaks on its carapace, as well as a mottled plastron.

All need similar care, with an aquaterrarium with a suitable basking site. On warm days, they can be transferred to an outdoor aquarium, with a rock so they can bask as well. A complete diet is essential to avoid nutritional problems.

REPRODUCTION Mature males have significantly longer claws than females, which they use as part of their display, fanning water in front of the female's face. Incubation may take 10 weeks or more, with the young measuring just over 1 in. (2.5 cm) when they hatch.

PAINTED TURTLE
An eastern painted turtle is shown here.

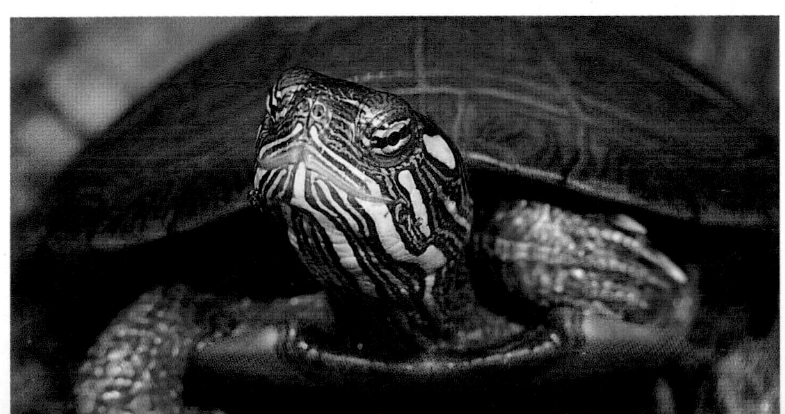

10 in. (25 cm)

Semi-aquatic

Terrestrial

Agree in groups
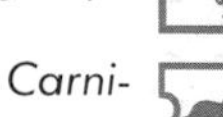
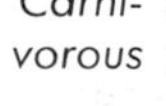

Carnivorous

6–20 eggs

RED-EARED TURTLE

Chrysemys scripta elegans

This species is found in an area from eastern USA to Mexico. It is the most widely kept turtle, with thousands being reared each year on farms specifically for the pet trade. In North America there are restrictions on the sale of small turtles, because of fears over *Salmonella* infection. Provided that several basic precautions are taken, these turtles should represent no danger whatsoever. Most important, wear gloves when cleaning out their quarters, and pour the dirty water down an outside drain rather than the household sink.

There are other sources of *Salmonella* bacteria in the home that are likely to be more dangerous, such as chicken carcasses and offal. It is not recommended to feed turtles on such foods, because they could become infected as a result. Nutritional problems are also likely if these reptiles are fed exclusively on meat. They are especially prone to swollen eyes, which are the result of a vitamin A deficiency, as well as soft shells, which may be linked with inadequate dietary calcium or a vitamin D_3 deficiency.

12 in. (30 cm)

Semi-aquatic

Terrestrial

Agree in groups

Carnivorous

8–12 eggs

Specially formulated diets, in the form of floating foodsticks prepared for these and other turtles, mean they can be reared satisfactorily without such worries. In addition, they will not soil the water in the same way as meat or fish, while a power filter can be used to remove the turtles' waste, so the tank will not need to be cleaned every day.

Young red-eared turtles have shells that are a relatively light shade of green, but this color is likely to darken significantly as they grow. It is important to realize from the outset that the cute little hatchlings will grow to a relatively large size, and will need substantially larger accommodations. Occasional amelanistic individuals have been recorded in the case of these turtles. Such individuals have a cream-colored carapace, with their red flashes on the sides of the face and the yellowish markings on their shell much as in the normal type.

Red-eared turtles are part of a genus whose distribution extends right down to South America. They are also sometimes described as "sliders," because of their habit of sliding into the water at any hint of danger. Young individuals can be especially nervous, but as they grow older they become much tamer. Avoid allowing them out into a backyard pond unless it is securely fenced, because there is a risk they could escape easily. In addition, larger individuals are likely to prey on any fish found there, as well as other aquatic life such as water snails.

REPRODUCTION Long front claws are a feature of males of this species. Courtship can be rather rough, with the male seeking to bite the female in order to gain a grip on her for mating purposes, although actual injuries are rare. Even if she lays her eggs in water, there is a possibility that they will hatch satisfactorily, although normally she will bury them on land, wetting the sides of the nesting chamber to stop it from collapsing. Hatching takes 13 weeks, and the young turtles may be mature from 4 years old.

RED-EARED TURTLE
Red-eared turtles must have an area of dry land in their quarters where they can bask. This is a young hatchling.

4 in. (10 cm)

Semi-aquatic

Terrestrial

Agree in groups

Carnivorous

2–5 eggs

EASTERN MUD TURTLE

Kinosternon subrubum

This species comes from eastern USA. Adults have brown shells, with yellowish underparts. They usually bask less than red-eared turtles, but should be kept in a similar way, with a secure partition between dry and wet areas of the vivarium. A suitable access point out of the water will also be required. Most aquarium shops can adapt a tank made for terrapins by placing a glass divider at one end to create a land area. This needs to be stuck in place with silicone sealant, as used in the seams of most modern glass tanks. A gently sloping ramp out of the water can be set in place in a similar fashion, but check that the edges of the glass are smooth so they cannot harm the turtle. Acrylic is less satisfactory because it can be scratched, and algae can then develop in the scratches.

EASTERN MUD TURTLE
The small size of these turtles makes them a good choice for breeding in a home vivarium.

REPRODUCTION To encourage spring breeding, it may help to overwinter adults at a lower temperature of around 50°F (10°C). It takes at least 5 years for young turtles to mature.

6 in. (15 cm)

Semi-aquatic

Terrestrial

Agree in groups

Carnivorous

1–5 eggs

COMMON MUSK TURTLE

Sternotherus odoratus

These small turtles are found in an area from southern Canada to south eastern Texas and Florida. They have evolved a particularly smelly way of protecting themselves, by releasing a volatile substance from skin glands. This has led to the species also being called "stinkpot," as well as "stinking Jim" in some parts of its range. In a vivarium, however, once they settle down, the turtles desist from this habit.

Common musk turtles are highly aquatic and primarily carnivorous in their feeding habits, although, like other turtles, they may sometimes eat aquatic vegetation. A balanced diet, as for other species, will keep them in good health, and avoids worries over possible nutritional deficiencies arising from excessive reliance on animal foods.

REPRODUCTION The tails of male musk turtles are generally thicker at their base and are longer overall. They may also have rough areas on the inside of their hind legs, helping them to clasp females during mating. Incubation lasts 15 weeks or longer. Hatchlings are about 1 in. (2.5 cm) in length. Maturity is reached once the turtles are about 3 in. (7.5 cm) long.

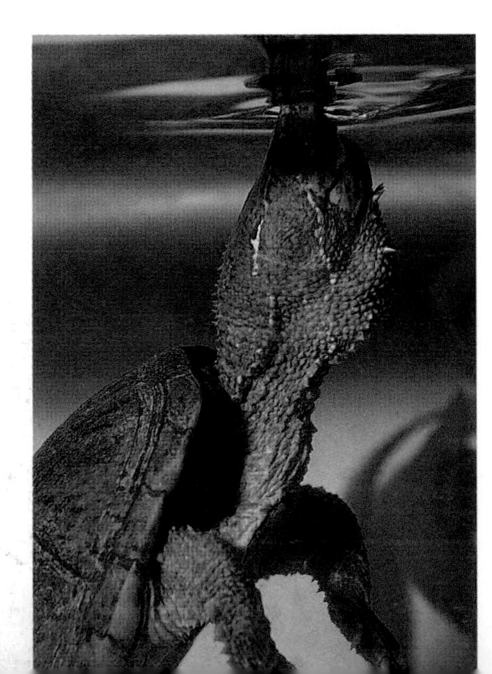

COMMON MUSK TURTLE
The musk turtle is another easily kept species.

FLORIDA SOFT-SHELL TURTLE

Trionyx ferox

20 in. (50 cm)

Aquatic

Aggressive

Carnivorous

10–22 eggs

Soft-shell turtles have a wide distribution, being found not only in North America, but also in parts of Africa and Asia. They all require similar care, although some species may grow to a much larger size than others. Soft-shells are almost entirely aquatic, and their tank should have a sandy bottom where the turtle can bury itself. The water level should be reasonably low, so that the soft-shell can extend its neck above the surface, while remaining on the bottom of the tank, with its long nostrils acting as a snorkel.

Aggressive by nature, young soft-shells are best kept separately. Otherwise there will be bullying, to the extent that the weaker individual does not eat properly, so that a difference in size between this turtle and its dominant companion starts to become apparent. The water in the vivarium for these turtles must be kept particularly clean. They are vulnerable to fungal infections, particularly until they are established in their quarters.

SOFT-SHELL TURTLE
All soft-shell turtles have leathery shells.

REPRODUCTION This is not feasible in an ordinary vivarium, as compatibility is often a problem. Females are usually larger. Their eggs are laid in sand, and incubation takes 3 months or less.

MATA-MATA

Chelus fimbriatus

16 in. (41 cm)

Aquatic

Aggressive

Carnivorous

12–28 eggs

Another highly aquatic chelonian, the mata-mata can make a fascinating vivarium occupant. They can be expensive to buy and feeding may be difficult. In the wild, in northern South America, the bizarre shape of these turtles helps to disguise them, as they lie on the bottom of the river surrounded by debris. When prey of a suitable size comes within range, the mata-mata sucks the unfortunate creature into its mouth by dropping its lower jaw.

Inert foodsticks are often refused by these turtles, which means that other foods, such as shrimp (sold for predatory fish by aquatic shops) are needed. Unfortunately, diseases can be introduced into the vivarium with such foods, so strain off the water beforehand as a protective measure. Small tropical fish will also be eaten, but it can be difficult to persuade these turtles to take safer prey, such as terrestrial invertebrates. Their tank should be equipped with a filter, and the water level should be kept relatively low, to allow the turtle to breathe from the bottom by extending its long neck.

MATA-MATA
The mata-mata is a very sluggish turtle, remaining inert for long periods.

REPRODUCTION Their eggs are likely to take nearly 30 weeks to hatch. Young mata-matas are a lighter shade of brown than adults, often set against a reddish-brown body color.

AMBOINA BOX TURTLE

Cuora amboinensis

There is often confusion between the Asiatic and New World box turtles, which differ significantly in their habits. The Amboina box turtle is one of the Asiatic species, some of which are aquatic, although they do roam on land as well. The name "box turtle" comes from the creatures' ability to seal themselves into their shell, using the movable flaps at the front and rear of the plastron for this purpose.

A vivarium for these turtles should consist primarily of water, although it should also incorporate an area of dry land. Hatchlings tend to spend more time in the water than bigger individuals. They prefer to feed in water rather than on land, taking foodsticks readily.

REPRODUCTION A male courts a potential mate by extending his head toward her and making exaggerated head movements, before snapping at her hind limbs to persuade her to stay still in order for mating to take place successfully.

7 in. (18 cm)

Semi-aquatic

Terrestrial

Agree in groups

Carnivorous

1–2 eggs

AMBOINA BOX TURTLE
The yellow stripe above the eyes and the dark carapace are characteristic of this species.

My turtles used to swim much more when they were younger, but now they are far less active. Is this normal? *Young turtles are naturally lively, and they rely on their ability to swim quickly to escape predators. Their shells give them little protection until they are bigger. Also, turtles vary in their level of activity. Some, such as the red-eared turtle or slider (see page 94), are avid swimmers, whereas others like the mata-mata (see page 97) rely on their inactivity as camouflage.*

COMMON BOX TURTLE

Terrapene carolina

Members of this species found in eastern USA vary quite widely in appearance. Some individuals have attractive patterns on their carapaces, whereas others have dull brown shells. Those which are brightly colored also tend to have more color on their front legs. In contrast to the Asiatic box turtles, the common box turtle is essentially terrestrial and it is often referred to as a tortoise in Europe.

A relatively shaded locality suits these turtles, with an outdoor run being suitable in the summer months in temperate climates when the weather is reasonably warm. A dish of water large enough for the turtle to submerge itself is essential. Feeding presents few problems. Foodsticks should form the basis of the diet, along with a variety of greens and fruit, with tomatoes often a particular favorite.

Live food such as mealworms can be offered. Outdoors, these turtles may catch worms, patrolling an area of lawn after a rain shower for this purpose. They can strike with considerable speed, and use their strength to pull the worm out of the ground. They also take snails, but such foods may carry parasites which could affect the turtle. Few chelonians are as long-lived as box turtles. There are reliable records of their living for well over 50 years, and possibly over a century.

What is the best way to tame my tortoise? *A young hatchling will be easier to tame than an adult. First, avoid sudden movements near your tortoise's head. This will cause it to pull back into its shell rapidly. Offer food with your fingers, and encourage your tortoise to feed from your hand in this way. You may then want to stroke the top of your tortoise's head. They often appear to like this!*

REPRODUCTION The difference in eye color is usually a reliable guide to the sex of the tortoise. Males have reddish-orange eyes, while females' eyes are browner. Males balance in a very upright position when mating, which takes place soon after the end of hibernation. Incubation takes 2–3 months on average.

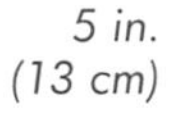
5 in. (13 cm)

Savannah

Terrestrial

Agree in groups

Carnivorous

4–7 eggs

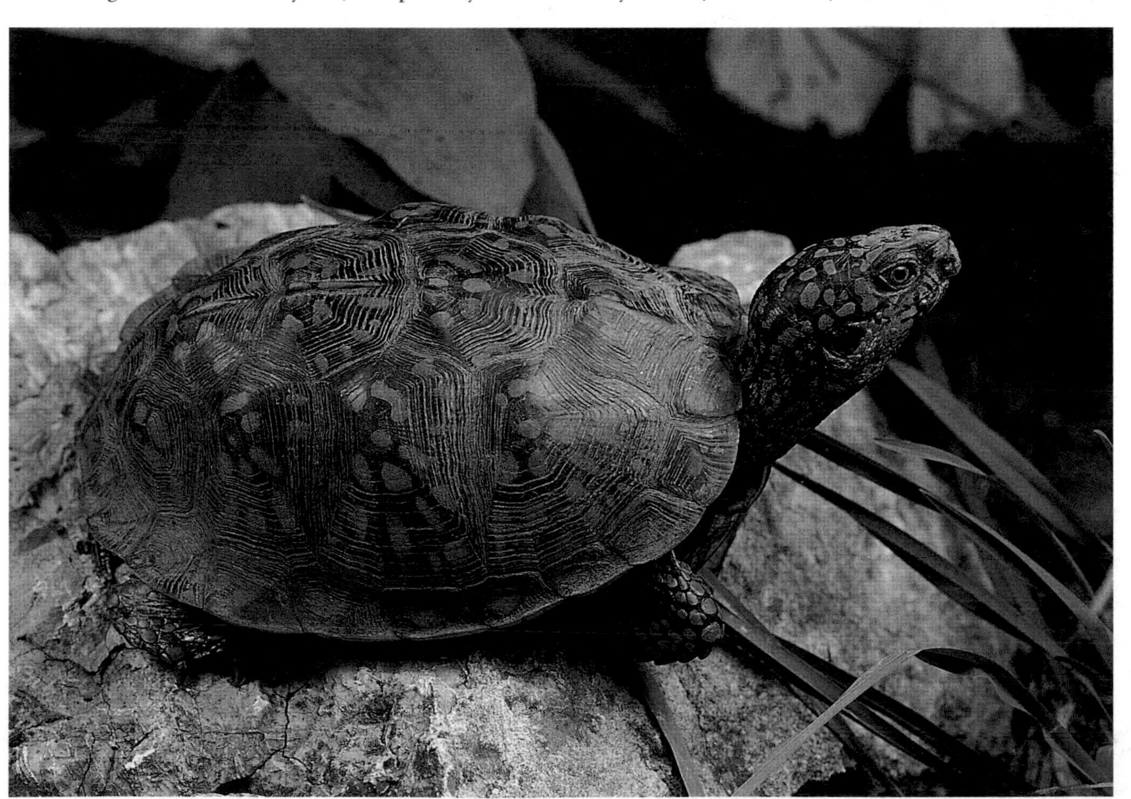

COMMON BOX TURTLE
Once box turtles are used to their environment, they lose their shyness.

AMPHIBIANS

There are two major amphibian groups of significance to the enthusiast, since the worm-like caecilians are not usually kept. They are the tailless frogs and toads, and the newts and salamanders, distinguishable by their characteristic long tails and narrow body shape.

There is no clear distinction between frogs and toads, although as a general guide frogs usually have smoother skins, longer legs and a more aquatic lifestyle than toads. Newts tend to develop pronounced crests running down the center of their backs, especially in the case of males, while this feature is not apparent in salamanders.

NEWT

FROG

SALAMANDER

Newts and salamanders are essentially terrestrial in their habits, but some frogs have evolved the ability to climb and live off the ground.

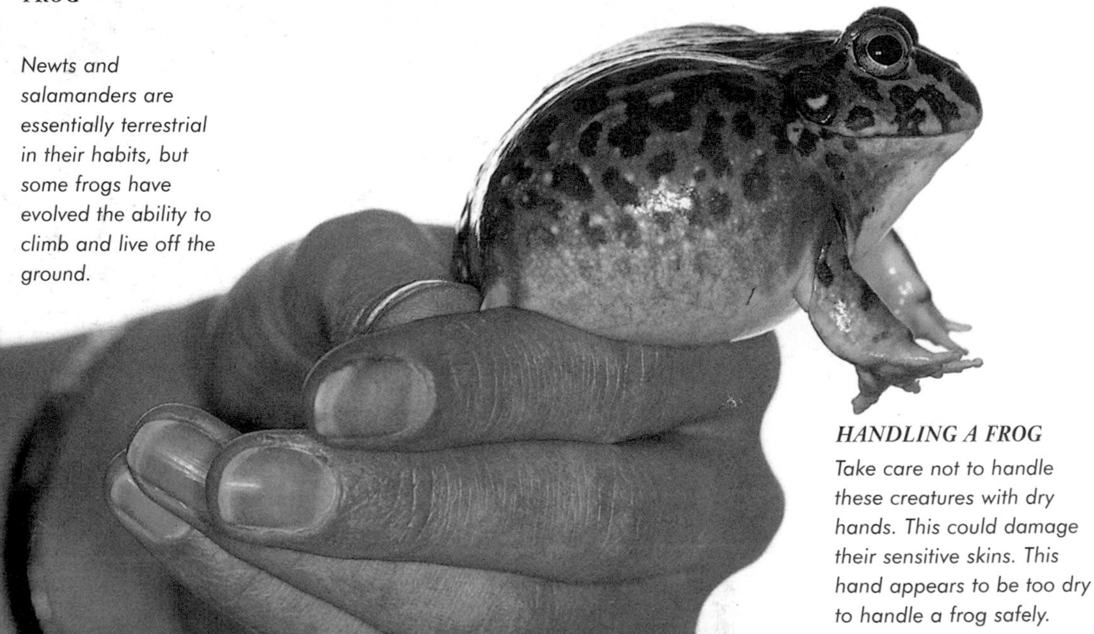

HANDLING A FROG

Take care not to handle these creatures with dry hands. This could damage their sensitive skins. This hand appears to be too dry to handle a frog safely.

Although amphibians may stray away from water, they will return to breed. Their spawn hatches into tadpoles that have gills for breathing purposes at this stage. These young amphibians will only develop lungs once they begin to metamorphose into adults.

To protect their sensitive skins, amphibians should be caught carefully with an aquarium net, from which they can be transferred to a suitable plastic carrying container with a ventilated lid. This should then be placed in a dark box, again with some ventilation, so that the amphibians will not be distressed by being permanently in bright light while in transit. Frogs and toads are far less likely to leap about in darkened surroundings, and so are at less risk of injuring their snouts.

Amphibians have a visual sense that is adapted to recognize prey by movement. As a result, they will generally ignore inanimate foodstuffs offered to them. If this type of food is moved in front of the creature with forceps, however, this will often encourage it to strike out and grab the morsel. Some amphibians can be quite vocal, especially in the case of male frogs and toads, which often call loudly and persistently to attract mates during the breeding period.

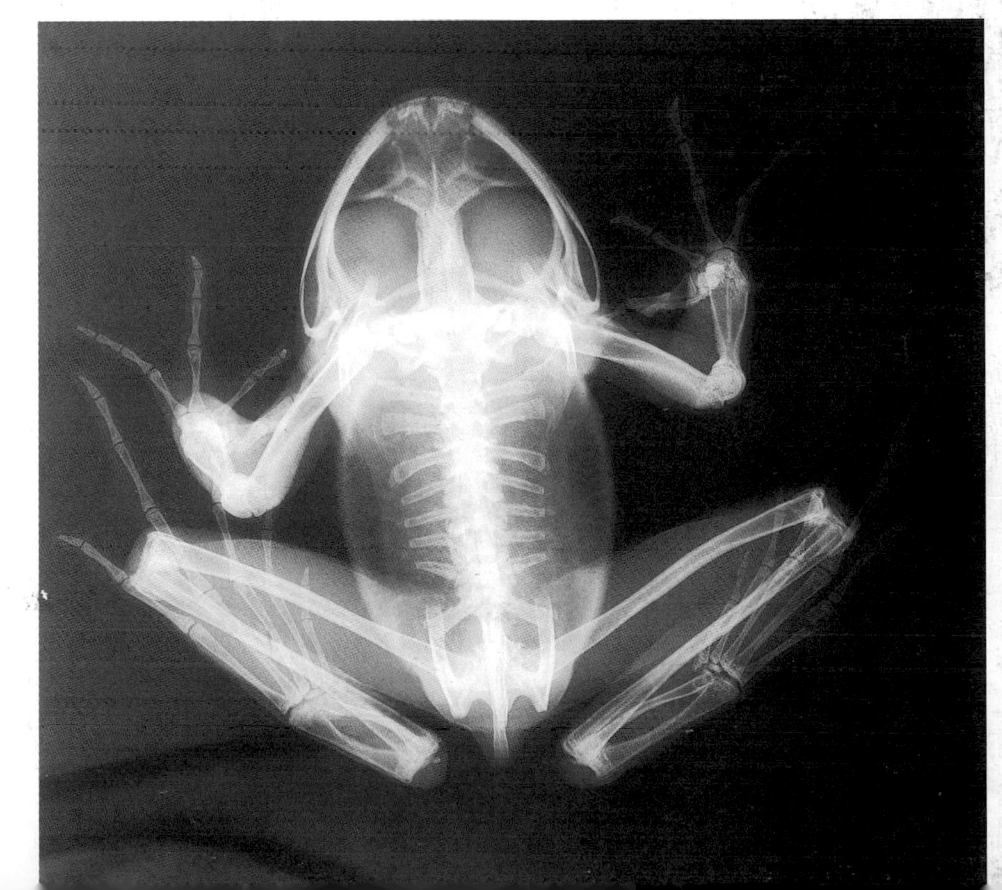

Amphibians have a skeleton not much different from that of higher vertebrates. Note the recognizable limb bones, ribs, and pelvis on this radiograph of a red-legged frog.

SURINAM TOAD

Pipa pipa

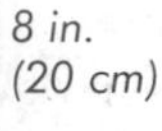
8 in. (20 cm)

Aquatic

May eat smaller companions

Carnivorous

100 eggs

These odd-looking toads from northern South America have a flattened body shape and a sharp, triangular face when viewed from above. They are highly aquatic, with strongly webbed hind feet, and need a spacious tank. Surinam toads also need heated water, and plants such as Java moss (*Vesticularia dubyana)* in their quarters, along with secure underwater hiding places.

SURINAM TOAD
Surinam toads can vary in color from gray to brown.

REPRODUCTION In the breeding season, females develop a ring-like swelling around the vent area. These toads are not easy to breed. Spawning takes place at the surface, and the male then directs the eggs onto the female's back. Within a day, the female's back swells up, with a pad embracing the eggs. Between 3 and 5 months later, the young toads break out. They can be reared on small live foods such as *Daphnia.*

DWARF CLAWED FROG

Hymenochirus boettgeri

1.5 in. (3.5 cm)

Semi-aquatic

Terrestrial

Agree in groups

Carnivorous

750–1,000 eggs

These aquatic frogs from Africa are relatively easy to look after in an aquarium with a sandy bottom. Suitable plants, and some structures such as special aquarium wood, which can break the surface, are also needed. Although dwarf clawed frogs are sometimes housed with fish, it is better to keep them alone. They may eat small fish.

DWARF CLAWED FROG
These small frogs are active swimmers.

REPRODUCTION Males can be recognized quite easily by special glands located just behind their front legs. This species will lay at any time of the year, with males in breeding condition calling to attract a mate. The eggs are again laid in batches at the surface of the water – but unfortunately they will be eaten rapidly by the adult frogs, so they should be scooped out with a tea strainer or similar utensil and transferred to a separate aquarium where they can hatch in safety. This takes place about 5 days later, with the young tadpoles digesting the remains of their yolk sacs for the next 2 or 3 days, before they become free-swimming. These tadpoles must be fed small pieces of meat from the outset. Metamorphosis takes nearly 2 months.

AFRICAN CLAWED FROG

Xenopus laevis

These frogs can be distinguished by the black claws on their inner toes. Their hind limbs are very powerful, helping them to swim either forward or backward. Their eyes are directed upward, to give good vision of their surroundings, since in the wild they spend much of their time underwater. The water level in their tank should be kept relatively low, as they breathe by standing up in the water. These frogs use their flexible long front toes to gather their food, which can consist of earthworms and similar invertebrates. It may be possible to wean them onto inert foods. The patterning on their backs varies, with some individuals being almost entirely grayish-black. There is also now a well-established albino strain of *Xenopus*, with a white body and red eyes.

AFRICAN CLAWED FROG
The flattened body shape of the African clawed frog is characteristic of aquatic frogs.

5 in. (13 cm)

Aquatic

May eat smaller companions

Carnivorous

1,000–2,000 eggs

REPRODUCTION African clawed frogs have been kept for many years in laboratories, where they were used for pregnancy testing until other methods were invented. Female frogs would spawn within 10 hours of an injection of urine from a pregnant woman. This method has since been adapted to achieve captive breeding, as otherwise these frogs do not spawn readily.

Males may be distinguished by their trilling calls, as well as the nuptial pads which develop in their front feet when they are in breeding condition. Plastic mesh must be fitted in the bottom of the spawning tank to prevent the frogs from consuming their spawn. They should then be transferred back to their aquarium, and the tadpoles will hatch within a couple of days. Fish fry rearing foods can be used as a first food, along with very fine pieces of egg yolk strained through cheesecloth. Some cannibalism will be unavoidable as the tadpoles grow.

Can amphibians climb glass? *Tree frogs, with their enlarged foot pads, are obviously well-adapted to climb around their quarters. Other amphibians may also be able to get out of a tank unless it is covered. Toads for example will often climb in a corner, pushing themselves up on the adjoining panels of glass using their hind feet, and then they grab onto the top, hopping out from here if the vivarium is uncovered.*

4 in. (10 cm)

Semi-aquatic

Terrestrial

Agree in groups

Carnivorous

400–1,000 eggs

EUROPEAN COMMON FROG

Rana temporaria

These frogs are found in a wide area from northern Europe to Central Asia. They tend to spend most of the year on land, in moist surroundings, returning to water in early spring after hibernation to spawn. A secure outdoor enclosure which meets these needs is the best way to house these frogs in temperate areas. They tend to have nervous natures and will not settle well in the confines of a vivarium, where jumping against the glass can damage their snouts. Feeding is straightforward, earthworms are a favored food. A suitable retreat for hibernation is also needed.

REPRODUCTION **Females will deposit huge masses of spawn in the pond, from which tadpoles hatch in a week or so, depending on temperature. The tadpoles will have completed their metamorphosis within weeks and will emerge onto land, hunting small invertebrates. They may live for 7–10 years.**

EUROPEAN COMMON FROG
Common frogs can vary in color from shades of brown through to green.

3.5 in. (9 cm)

Semi-aquatic

Terrestrial

May eat smaller companions

Carnivorous

400–1,000 eggs

LEOPARD FROG

Rana pipens

There is some confusion over the classification of these frogs, as several different species are described under this name. They are found in an area from southern Canada to Mexico. The patterning on their skins has given rise to their common name. Provide them with a vivarium with suitable retreats and a soft substrate complete with some sphagnum moss. The frogs are likely to burrow here. A water bowl is also essential. Leopard frogs are predatory and will feed on invertebrates, such as crickets.

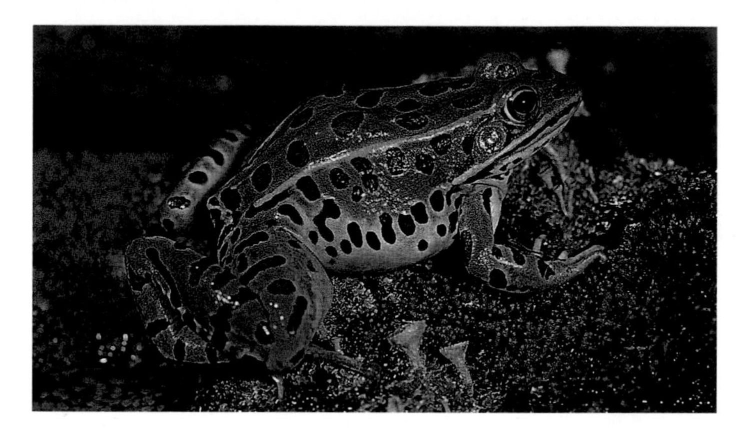

LEOPARD FROG
These frogs vary in markings through their wide range.

REPRODUCTION **Spawning takes place in the spring, when the frogs should be transferred to an aquarium with water weed such as Canadian pondweed (traditionally known as *Elodea canadensis*). Hatching normally takes 2–3 weeks at a water temperature of 50°F (10°C), and after 4 months the tadpoles will have changed to young leopard frogs. At this stage they will be nearly 1 in. (2.5 cm) long.**

ORNATE HORNED FROG

Ceratophrys ornata

5 in. (12 cm) 

Tropical woodland 

Terrestrial 

Aggressive 

Carnivorous 

1,000 eggs 

These Argentinian frogs are highly predatory. They eat invertebrates, and other creatures, including pinkies, as they grow larger. Horned frogs spend much of their time buried in the substrate of their vivarium. A thick layer of bark chips or similar material should be provided for this purpose. A large water dish and sphagnum moss, which can be sprayed to increase the relative humidity, should also be available.

REPRODUCTION Mature males have darker markings on their throats, and grow nuptial pads on their legs when they are in breeding condition. Cooling their quarters down to 68°F (20°C) for up to 3 months, and then running water over them in warmer surroundings, may stimulate breeding. Hatching takes place in a day or less. The young should be reared in small groups to reduce cannibalism.

ORNATE HORNED FROG
This frog has a swollen area above each eye, which are its "horns."

PAINTED FROG

Kaloula pulchra

3 in. (7.5 cm) 

Semi-aquatic 

Terrestrial 

May eat smaller companions 

Carnivorous 

500 eggs 

This species originates from south east Asia and is also known as the Asian bullfrog and as the Malayan narrow-mouthed toad. These frogs are reasonably easy to keep in a vivarium lined with soft material into which they can burrow. Their small mouths mean that they can only tackle fairly small invertebrates, but they have healthy appetites, so their quarters will need frequent cleaning.

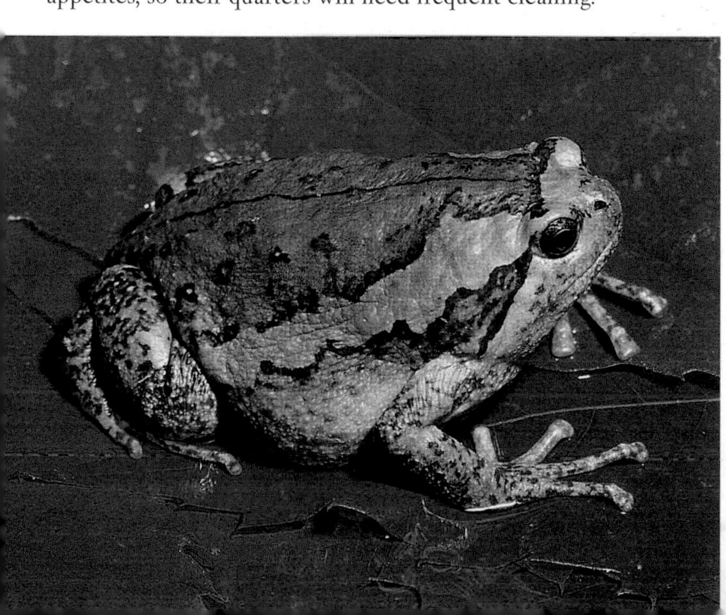

REPRODUCTION Pairs can be recognized quite easily, because males have dark throats and are smaller than females. Breeding in the wild takes place during torrential monsoons, so allow their quarters to become drier over the course of several weeks, then spray daily and increase the amount of water that is available. These frogs normally breed in temporary pools of water, so metamorphosis is very quick, lasting just 2 weeks.

PAINTED FROG
The broad tan stripes running across the head and down the sides of the body are characteristic of the painted frog.

2.5 in.
(6 cm)

Tropical woodland

Arboreal

Agree in groups

Insectivorous

200 eggs

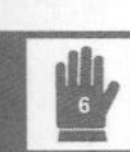

MARBLED FROG

Scaphiophryne marmorata

A number of Madagascar's anurans are now being kept and bred successfully. They are unique, with no close relationship to frogs and toads found in mainland Africa. The marbled frog is, in fact, the only member of its genus. Its basic coloration is olive-green, broken by darker markings, but there can be considerable differences in patterning between individuals. The "marbling" helps to provide camouflage in the forested areas of the island. As they are capable of climbing, they require a tall vivarium, which can include natural or artificial plants. Small invertebrates should form the basis of their diet.

REPRODUCTION Virtually nothing is known about the breeding habits of these tree frogs. Presumably, their requirements are similar to those of other members of their family, the Microhylidae.

MARBLED FROG
The disks on the feet of these frogs act as suction pads.

4 in.
(10 cm)

Tropical woodland

Terrestrial

May eat smaller companions

Carnivorous

1,000 eggs

SOUTHERN TOMATO FROG

Dyscophus guineti

These frogs, also from Madagascar, have become very popular, thanks in part to their highly unusual coloration. They were almost unknown until the 1980s, and there is still dispute about their taxonomy. Scientists are uncertain as to how many true species exist. Tomato frogs have proved quite straightforward to keep, although they will burrow and may hide away for long periods in their vivarium, venturing out to feed at intervals. They eat a variety of invertebrates.

REPRODUCTION Female tomato frogs are significantly larger than males. It is possible to persuade them to spawn naturally by reducing the humidity in their quarters for 4 weeks or so, and then raising it again, to mimic the effects of heavy rainfall. The eggs hatch rapidly and can be reared on flake food, as sold for fish. Tadpoles will complete their transformation from the age of 6 weeks onward. It is essential to give them a ramp to allow them to climb out of the water. Tomato frogs mature rapidly, and may breed at a year old.

SOUTHERN TOMATO FROG
Tomato frogs rank among the most highly prized anurans.

GOLDEN MANTELLA

Mantella aurantiaca

These small frogs are just one of a number of species of highly colored frogs originating from Madagascar. They all appear to need similar care, with plenty of retreats in their vivarium being essential. Damp moss provides a suitable carpet for the floor covering of their quarters, and wood should be used to provide further decor. Small invertebrates such as crickets, sprinkled with a nutritional balancer, may be offered as food. An area of water, concealed in a dark place, is also important. Their quarters should be kept humid. These frogs should be handled with care, as toxins may be present in the skin.

REPRODUCTION Keeping a small group of these frogs together should help to ensure that you have at least one pair, although males may be recognized on occasions by their calls. Lower the temperature to 72°F (22°C) for 2–3 weeks, and then raise the humidity. Females will soon swell with eggs, which they will ultimately lay in secluded damp parts of the vivarium. Both sexes attend the eggs to prevent them from drying out. Hatching occurs 2–3 months later, and the tadpoles should be reared individually. The eggs are likely to be concealed under moss or elsewhere in the vivarium.

GOLDEN MANTELLA
Golden mantellas can range in color from yellowish-orange to a fiery orange shade.

1.5 in. (3 cm)

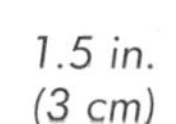

Tropical woodland

Terrestrial

Agree in groups

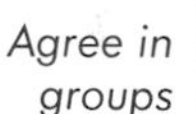

Insectivorous

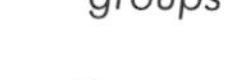

20–30 eggs

POISON ARROW FROGS

family Dendrobatidae

This highly-colorful group of frogs from Central and South America includes the most deadly species known on Earth, which is referred to by its scientific name of *Phyllobates terribilis* – it has no common name. The skin of this dull yellow frog contains enough poison to kill up to 10 people. South American tribespeople extract the venom by roasting the frogs over a fire, and tipping their arrows with the resulting outflow of deadly liquid.

Under normal circumstances, however, with careful handling only when necessary, there is no threat to their owner's health. Note that if you have any cuts on your hands you must wear disposable gloves, to prevent any venom from entering your body. Otherwise, you may soon start to feel distinctly unwell. Interestingly, as these frogs have been bred to several generations, the potency of their venom has declined, presumably because it is not necessary for their survival in these surroundings.

Bright colors in nature tend to indicate danger, which is certainly true in the case of these frogs. They can vary from a stunning azure blue with black to an equally striking red and black. Individual variation is unusually wide, to the extent that no one is certain how many species should be recognized, although around 65 is a realistic assessment.

Poison arrow frogs occupy a range of ecological niches in the rainforest environment. A few are arboreal, while others occupy the forest floor, concealing themselves under leaf litter and other debris. Some poison arrow frogs live in the vicinity of streams running through the forest. Their quarters need to be well planted, and small invertebrates should be provided as food.

1.5–2 in. (3–5 cm)

Tropical woodland

Arboreal

Agree in groups

Insectivorous

5–10 eggs

POISON ARROW FROGS
The markings of the red and black poison arrow frog (Dendrobates histrionicus) are among the most variable.

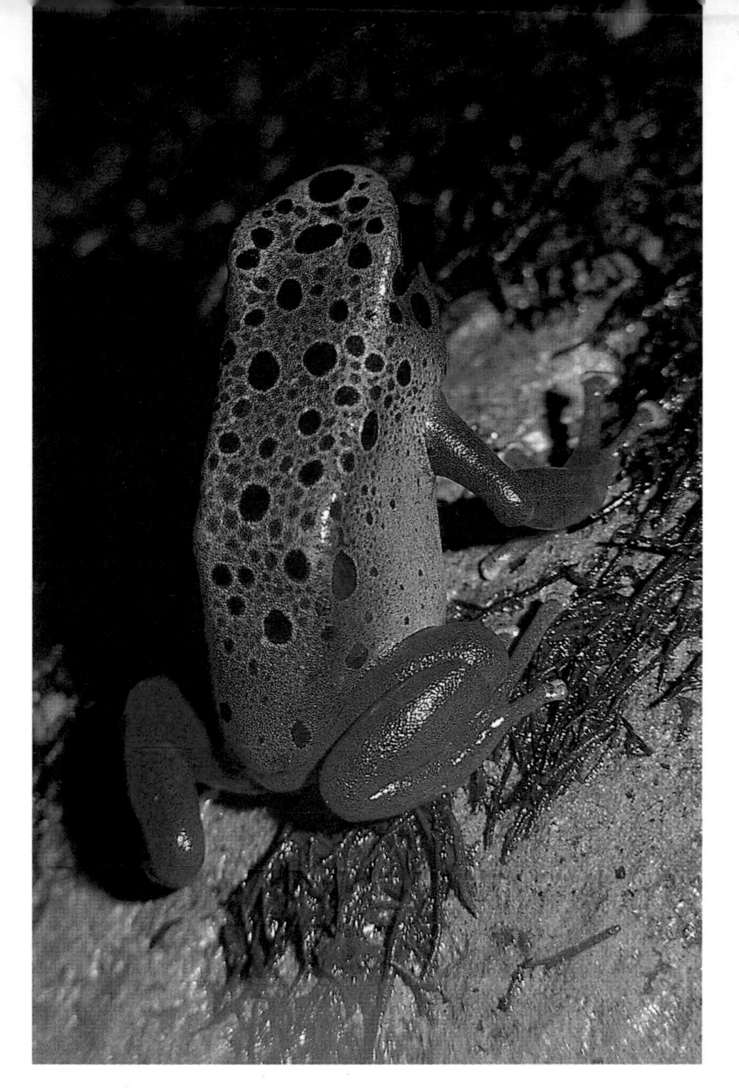

POISON ARROW FROGS
*The blue poison arrow frog (*Dendrobates azureus*) is one of the species now being bred successfully and regularly.*

REPRODUCTION Living in places where there is not much water, poison arrow frogs have evolved a breeding strategy very different from that of most other anurans. Their eggs are often laid on the ground, with the frogs remaining nearby. Once the eggs hatch, the male, distinguishable by his thicker second and third front toes, carries the tadpoles to water, so they can metamorphose into frogs. The male frog may, for example, put the eggs in the cup formed by a bromeliad's leaves, where rainwater is easily trapped.

Food may not be available in such places. Remarkably, some female poison arrow frogs actually care for their offspring, by laying infertile eggs which the young can eat. This was only discovered thanks to the keen observation of breeders in 1980. It is clear that eggs should be left in place in the vivarium. In an emergency, egg yolk rubbed through cheesecloth will suffice as a rearing food for the tadpoles, although their growth rate will be slowed significantly. While they would normally emerge as frogs after a period of 2 months, this may be extended to 6 months if they are fed on egg yolk.

2 in. (5 cm)

Temperate woodland

Arboreal

May eat smaller companions

Insectivorous

200 eggs

EURASIAN GREEN TREE FROG

Hyla arborea

Swollen toe pads are characteristic of tree frogs generally, helping them to hold onto vertical surfaces without losing their grip. This particular tree frog originating from southern Europe and western Asia, has been bred successfully for many years, and is ideal for a greenhouse or similar escape-proof environment, where there is shade available during the summer. The frogs should not be exposed to temperatures much above 77°F (25°C). It is advisable to bring them indoors in the winter months, when a large unheated vivarium, equipped with an ultraviolet light should be provided for them.

REPRODUCTION A shallow pond with water plants should be provided for spawning. Males compete with each other in song. The winning male will mount the female and together they will enter the water, where spawning takes place. It will take roughly 2 months for the tadpoles to change into young green tree frogs.

EURASIAN GREEN TREE FROG
Humidity is important in the care of tree frogs. Their quarters should be sprayed regularly.

1.5–2 in. (3.5–5 cm)

Temperate woodland

Arboreal

May eat smaller companions

Insectivorous

700 eggs

AMERICAN GREEN TREE FROG

Hyla cinerea

This species from central and eastern USA can be distinguished from its European cousin by its more slender shape and usually by the cream stripe running down each side of its body, although this may not be present in every case. American green tree frogs also have slightly larger toe pads. They are less adaptable in terms of temperature, with 72–75°F (22–24°C) being ideal during the daytime. High humidity is important, and there must be plenty of tall plants in their vivarium. They will eat various invertebrates, including crickets.

REPRODUCTION Males call loudly, inflating the sac under their throat when vocalizing, with such calls reaching a peak during the breeding season. Cooling their enclosure down to 50°F (10°C) in the late winter for several weeks is likely to trigger breeding activity in the spring, but obviously this is only to be recommended for fit adult tree frogs which have been feeding well beforehand. Provide shallow water with plants for spawning. Hatching takes 6 days, and the young frogs will leave the water about 2 months later.

AMERICAN GREEN TREE FROG
This is one of the most widely kept tree frogs.

AMERICAN GRAY TREE FROG

Hyla versicolor

2 in.
(5 cm)

Temperate woodland

Arboreal

May eat smaller companions

Insectivorous

2,000 eggs

The unusual gray patterning of these tree frogs conceals them when they are resting on the bark of trees. Gray tree frogs also have orange markings on the inner surface of their thighs.

Their care is straightforward and does not differ significantly from that of their green relatives. A tall and spacious vivarium is essential, which should include pieces of cork bark set vertically for climbing, in addition to plants.

REPRODUCTION This species is not commonly bred, but this could be achieved without too much difficulty by treating it like the American green tree frog (see page 110).

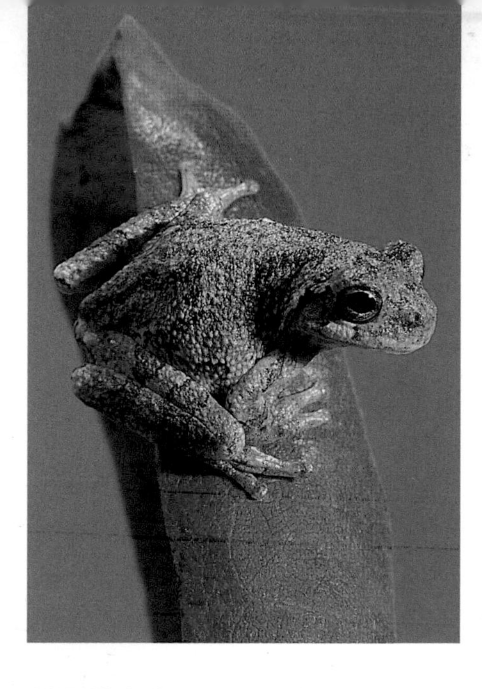

AMERICAN GRAY TREE FROG
A cream spot beneath the eye helps to distinguish this species.

RED-EYED TREE FROG

Agalychnis moreletii

3 in.
(7.5 cm)

Semi-aquatic

Arboreal

May eat smaller companions

Insectivorous

75 eggs

Prominent red eyes set this species from Central America apart from other mainly green tree frogs. Creamy barring on the flanks, along with the yellowish underparts, help to identify this particular red-eyed tree frog. There are eight different species in total. Unfortunately, it is often difficult to appreciate their natural beauty, because red-eyed tree frogs are primarily nocturnal. During the day they hide under leaves, or in similar places. These phyllomedusine frogs require a very high humidity, and a small waterfall operated by an aquarium pump is beneficial. The temperature in their quarters should not fall significantly at night, because this will deter them from feeding. Invertebrates such as crickets, which will jump readily, are an ideal food, provided they do not attack the vegetation in the vivarium.

RED-EYED TREE FROG
The spectacular appearance of these tree frogs is enhanced by their orange toe pads.

REPRODUCTION Males are smaller than females. Lowering the humidity briefly should encourage breeding. Eggs are laid above the water on overhanging vegetation. The tadpoles wriggle down into the water after hatching. Fish food makes a good rearing food, and natural light can be beneficial. The young need an easy way out of the water while they develop. They will start breeding in about 2 years.

2 in.
(5 cm)

Semi-aquatic

Terrestrial

Agree in groups

Carnivorous

600 eggs

RED-BANDED CREVICE CREEPER

Phrynomerus bifasciatus

The attractive red markings are a warning of toxins in the frog's skin. They may change color according to their environment, becoming more orange under certain circumstances. They are found in parts of tropical and southern Africa.

These frogs are often found some distance from water, hiding under rocks. They rely on concealment to avoid danger, because they are not good jumpers. They emerge mainly at night to hunt for termites and ants, although they will eat crickets as a substitute in a vivarium. The substrate should be relatively deep, with adequate hiding places, and not too dry.

REPRODUCTION These frogs have no set breeding period in the wild, they take advantage of the heavy rains when they come. Having been kept in relatively dry surroundings for a month or so, the frogs should be moved to a shallow aquarium. After a week, the addition of more dechlorinated water should cause spawning. The frogs can then be removed and the eggs will hatch within 4 days. Metamorphosis is completed in 2 months.

RED-BANDED CREVICE CREEPER
This species is also known as the rubber frog.

4.5 in.
(11.5 cm)

Tropical woodland

Arboreal

May eat smaller companions

Carnivorous

150 eggs

WHITE'S TREE FROG

Litoria caerulea

These dumpy tree frogs from New Guinea and north eastern Australia have placid natures and great personalities, which has helped to ensure their popularity. They are easy to keep in a tall vivarium, with plants which have tough leaves and stout stems to support the weight of this relatively large species. Larger invertebrates should be offered as food. A dish of water should be available on the floor, although a larger area of water is necessary for breeding purposes. Their color can vary from green to shades of blue, and piebald individuals occur occasionally, with variable areas of white instead of green.

REPRODUCTION Males are smaller, with darker throats, and call frequently during the breeding period. Hormonally induced spawning is common, but it is possible to persuade them to breed by cooling and reducing the humidity in their vivarium, before spraying regularly twice a day to mimic rainfall, and raising the temperature again. Spawning should begin within a week. Good natural lighting is essential in order to develop the color of the young as they metamorphose. This will probably take at least a month from egg-laying.

WHITE'S TREE FROG
White's tree frog can live for 15 years or more.

ASIAN TREE FROG

Polypedates leucomystax

This species from India and south east Asia is also known as the white-bearded flying frog and given an alternative scientific name, *Rhacophorus leucomystax.* It belongs to a group of frogs which live in trees and are very powerful jumpers, able to leap from branch to branch with little difficulty. They have the rounded, expanded toe pads characteristic of tree frogs, and need to be housed in a similar way. Insects which venture up to the higher levels of the vivarium are most likely to be eaten.

REPRODUCTION Males are smaller, and have thumb pads which grow in the breeding period. Branches overhanging water are important if this species is to breed in a vivarium. Eggs are in a ball of foam, which includes the male's sperm. The hatched tadpoles drop into the water about 5 days later. They can then be reared separately, with young froglets emerging onto land about 4 weeks later.

ASIAN TREE FROG
Simulating heavy rainfall is an important breeding condition for the Asian tree frog in vivarium surroundings.

3 in. (7.5 cm)

Tropical woodland

Arboreal

May eat smaller companions

Insectivorous

20–60 eggs

COUCH'S SPADEFOOT TOAD

Scaphiopus couchii

Spadefoot toads are to be found in both North America and parts of Europe. This American species, like its New World relatives, requires significantly drier surroundings than European species. Spadefoots have a projection on each leg which allows them to dig into the ground very quickly to escape danger. The floor covering of their vivarium should consist of a thick layer of fine grit, in which the toads will hide, emerging at dusk. They can tap moisture hidden deep in the soil, to prevent themselves from becoming desiccated.

COUCH'S SPADEFOOT TOAD
Couch's spadefoot has the most striking markings of this group of toads.

REPRODUCTION It is hard to achieve this in the confines of a vivarium, but moving the toads to a greenhouse, running a hose over part of their accommodation, and setting small plastic containers filled with water in the ground here may work. In the wild, these toads breed after heavy rain, taking advantage of temporary pools which form.

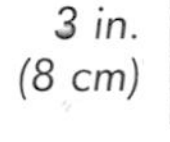

3 in. (8 cm)

Savannah

Terrestrial

May eat smaller companions

Insectivorous

200–250 eggs

COMMON EUROPEAN TOAD

6 in. (15 cm)

Temperate woodland

Terrestrial

May eat smaller companions

Carnivorous

2,000–10,000 eggs

Bufo bufo

The coloration of these toads varies through their wide range, from shades of reddish-brown, which is quite unusual, through to dark brown with exceptionally prominent "warts" or skin swellings. Found in Europe, Asia, and North Africa, common European toads will thrive in an outdoor vivarium in temperate areas, becoming most active after rain. They should have retreats in their enclosure and require a varied diet of invertebrates to keep them in good health. Their appetite for garden pests such as slugs means that they can be an asset if allowed in an organic vegetable garden.

REPRODUCTION **Breeding takes place in the spring, with females laying strands of eggs draped around pond plants. Males are smaller and display nuptial pads that are especially prominent at this time of year. The young toads will emerge onto land when they are between 2 and 3 months old. Small live foods, such as fruit flies, will be needed for them at this stage, and cultures of these should be prepared well in advance.**

COMMON EUROPEAN TOAD
Copper-colored eyes are a feature of the common European toad.

AMERICAN TOAD

3.5 in. (9 cm)

Temperate woodland

Terrestrial

May eat smaller companions

Carnivorous

4,000–8,000 eggs

Bufo americanus

Another adaptable species with a wide range, the American toad occurs in southern Canada and over central and eastern USA, in a wide variety of habitats, only returning to areas of water for breeding. The parotid glands, located on each side of the body at the back of the neck, are relatively prominent, as are the raised areas of skin on the head. It can sometimes be confused with Woodhouse's toad (*Bufo woodhousei*), but American toads have only one or sometimes two wart-like swellings in the dark markings on their backs. The care of these toads is similar to that of the previous species, although it has a reputation for being shyer, so it needs more retreats in its quarters. Broken flowerpots set on their sides are ideal for this purpose.

REPRODUCTION **Breeding follows the typical bufonid pattern, with the males entering the water in the spring and calling for mates. A fairly large area of water is necessary for spawning.**

AMERICAN TOAD
American toads appreciate rather moist surroundings.

GREEN TOAD

Bufo viridis

GREEN TOAD *Having relatively short hind legs, green toads tend to walk rather than hop.*

6 in. (15 cm)

Temperate woodland

Terrestrial

May eat smaller companions

Carnivorous

12,000–18,000 eggs

Another temperate species, the green toad usually has very attractive patterning, although there are noticeable differences through its range. Individuals originating from northern Africa and from eastern parts of Asia tend to have a more yellowish-brown appearance, which may help them to blend in more successfully in a sandy environment. Green toads can be maintained in a secure outdoor vivarium for the warmer months, where breeding success is more likely to be achieved.

REPRODUCTION Males are smaller in size and have slightly wrinkled skin covering their throat, which expands into a vocal sac when they start to croak. Hibernation at 39°F (4°C) – possibly slightly higher for southerly subspecies – is a necessary prelude to breeding in the spring, and huge numbers of eggs are produced in the summer. Only a few of the young tadpoles will survive, with cannibalism common. Plenty of small invertebrates will be necessary for the young toads once they come onto land.

AMERICAN GREEN TOAD

Bufo debilis

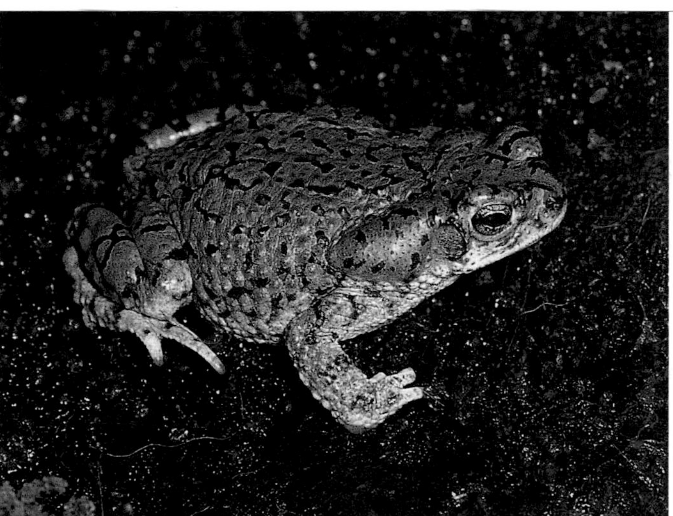

2 in. (4.6 cm)

Temperate woodland

Terrestrial

Agree in groups

Insecti-vorous

150 eggs

The requirements of the American green toad are similar to those of its European relative, but its smaller size makes it perhaps easier to cater for, particularly during the breeding season, when most bufonids must be provided with relatively large areas of water for spawning. It tends to have a shyer nature, however, and is thought to be mainly nocturnal in its habits. A vivarium for American green toads should incorporate a sandy area, along with areas of moss kept damp by regular spraying in which the toads can hide. Other hiding places can also be incorporated into their quarters.

REPRODUCTION As with other toads from temperate parts of the world, lowering the temperature in their quarters during winter is essential for the onset of breeding during the following spring. Males have darker throats and develop nuptial pads. Metamorphosis of the tadpoles may take nearly 2 months.

AMERICAN GREEN TOAD *An attractive combination of lime green with black markings characterizes this species.*

3 in. (7.5 cm)

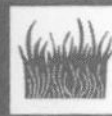
Savannah

Terrestrial

May eat smaller companions

Insectivorous

2,000 eggs

RED-SPOTTED TOAD

Bufo punctatus

These toads originate from south western USA to Mexico in arid surroundings. They are very agile, and as with other jumping species, their vivarium should be covered with a ventilated hood. Moist sand and rockwork will recreate their natural environment, with some damp moss for them to burrow in. A suitable bowl of water should be provided, although they will not stay immersed here for long periods. Feeding presents few problems, with a variety of invertebrates being taken readily, including wax-moth larvae and mealworms, although the latter may prove more indigestible. During the winter, the toads will burrow underground, emerging again in the spring. They may be toxic if swallowed by pets or humans.

RED-SPOTTED TOAD
Red spots are widely spread over the body of these toads.

REPRODUCTION Males make a sound not unlike that of a singing bird when they are calling, but relatively little has been recorded about their breeding habits. It is unlikely to differ substantially from that of other members of the genus. Spawning in the wild has been recorded in shallow pools, where these toads congregate.

1 in. (2.5 cm)

Temperate woodland

Terrestrial

Agree in groups

Insectivorous

500–800 eggs

OAK TOAD

Bufo quercicus

One of the smaller North American species, the oak toad has a surprisingly loud call for its size. It normally lives in fairly dry surroundings, mostly woodland and grassland areas. The vivarium can be lined with bark chippings and rocks as well as some plants to create an attractive backdrop and provide the toads with some cover. Invertebrates such as young crickets make a suitable food, treated with a balancer to correct their nutritional deficiencies beforehand.

REPRODUCTION These toads appear to breed in the summer whenever conditions are favorable – that is, when there are areas of water after heavy rains. It has proved possible to encourage spawning by keeping the toads well fed and then simply moving them to an aquarium containing aquatic plants and several inches of dechlorinated water. Females will lay their eggs in huge numbers of individual strands, each of which measure about 2 in. (5 cm) long, and contain around 10 eggs each. Fruit flies and aphids will be greedily consumed by the young toads as they emerge onto land.

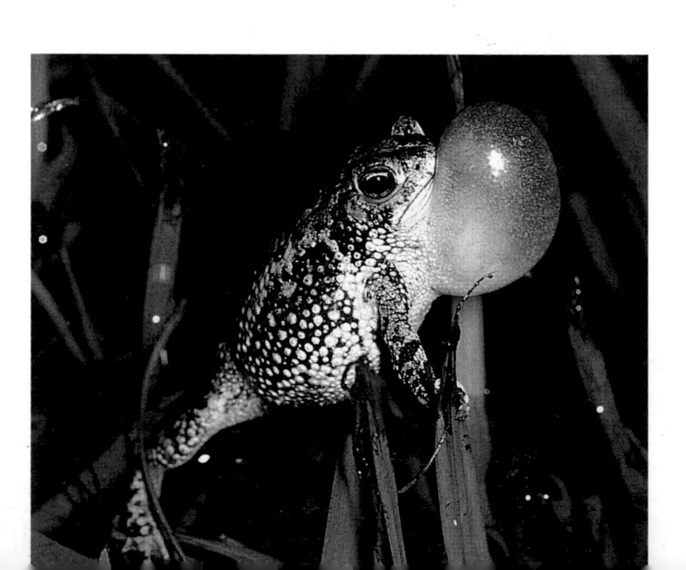

OAK TOAD
A white stripe down the center of the back (not visible in picture) is characteristic of the oak toad.

GIANT TOAD

Bufo marinus

One of the most notorious toads, this species is loved and hated by people in equal measure. It was deliberately introduced into Australia, in the hope that it would control sugarcane beetles. Unfortunately, it failed and has since developed a taste for many of Australia's smaller creatures, including fellow amphibians, and has itself become a serious pest. Giant toads have also established themselves on other islands, including Hawaii and New Guinea, with similarly devastating results.

In some cases, however, people keep these toads as household pets. Remarkably, they will often settle down well in such surroundings, although they will not necessarily agree with other pets. If a dog tries to pick up one of these toads in its mouth, it will soon start to salivate profusely and may suffer other more serious side effects, because of the deadly bufotoxin present in the prominent parotid glands on the sides of the toad's neck. Relatively dry surroundings are essential. These toads have large appetites, and will eat larger invertebrates and pinkies. They may even take dog and cat food.

REPRODUCTION Females generally grow to a larger size than males, but captive breeding is unlikely to be successful in anything other than a large greenhouse-type structure. Tadpoles develop quickly, changing into young toads within a month. At first they have a variegated pattern, becoming darker as they mature.

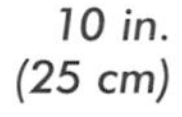
10 in. (25 cm)

Tropical woodland

Terrestrial

Aggressive

Carnivorous

20,000–30,000 eggs

GIANT TOAD
A character, but also a handful, this is the largest of all toads. Sometimes also known as the cane toad.

Is it possible to tame a toad? *Some toads can become very tame, particularly if they are fed regularly by hand. Start by offering food on a pair of blunt-ended forceps. Then once the toad is feeding readily in this way, you can dispense with the forceps, and feed it directly by hand. The cane toad (see above) is a good choice as a toad to tame because it is large in size, and quite bold by nature.*

ORIENTAL FIRE-BELLIED TOAD

Bombina orientalis

Being highly attractive and easy to care for, these toads are popular. They are found in north east China, Korea, and the former U.S.S.R. They can be distinguished from other fire-bellied toads by the coloration of their underparts, which are scarlet broken by black markings. They need a vivarium which incorporates enough water for swimming, with hardy aquatic plants.

The land area should include sufficient retreats for these frogs to hide, although once they are established in their quarters, they are not shy and will even come to feed from the hand. A variety of invertebrate foods, such as mealworms and wax-moth lavae, will be taken in this fashion. Oriental fire-bellied toads will not usually prey on aquatic invertebrates, preferring to feed on land, although they may also take creatures from the water surface. They are toxic if swallowed by pets or humans.

2 in. (4.5 cm)

Semi-aquatic

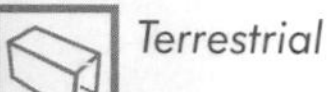
Terrestrial

Agree in groups

Insectivorous

300 eggs

REPRODUCTION Cooling their quarters down to 50°F (10°C) in the late winter should stimulate breeding in the following year. It is important to encourage these toads to eat well in the late summer, so that they are well conditioned for their period of relative inactivity. Males will start to call, often after dark, from late spring. They can be clearly identified by their nuptial pads at this stage, while females swell with eggs.

The brilliant scarlet coloration tends to be dull in captive-bred stock unless they have been reared on live foods which contain plenty of the coloring agent carotene, such as aphids and daphnia. Another method is to use powdered coloring agents produced for "color feeding" certain birds, sprinkling tiny quantities over the food of the young toads. This is best accomplished by placing the live food in a plastic bag and shaking it with the powder, which should stick to it.

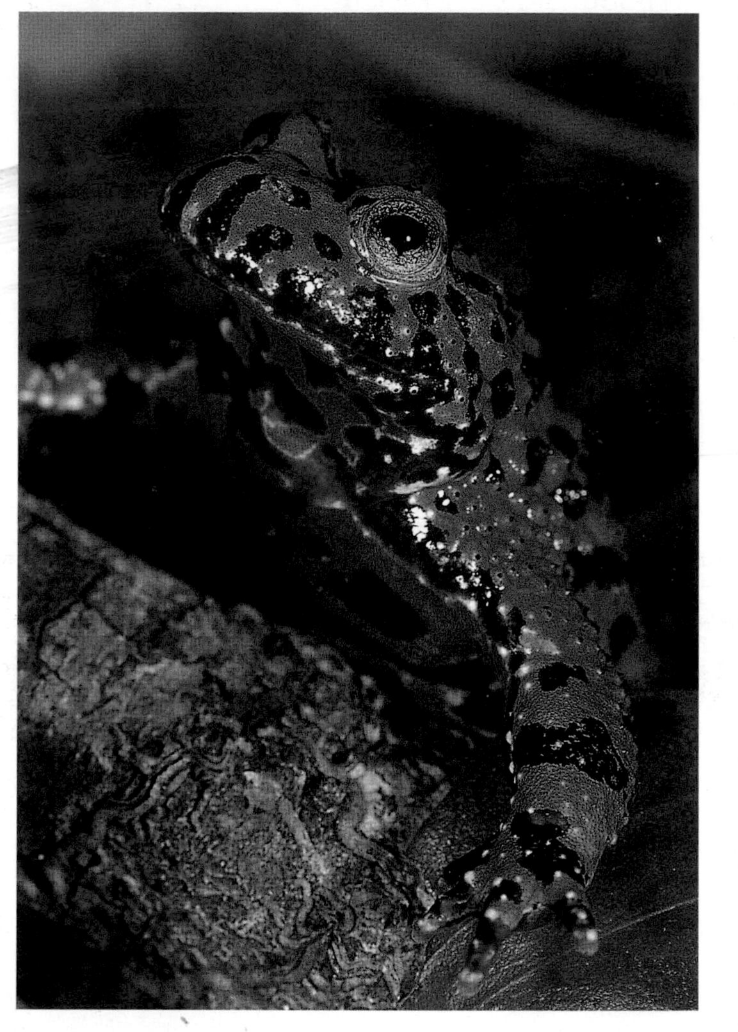

ORIENTAL FIRE-BELLIED TOAD
Oriental fire-bellied toads can be mature by the time they are a year old.

EUROPEAN FIRE-BELLIED TOAD

Bombina bombina

2 in.
(4.5 cm)

Semi-aquatic

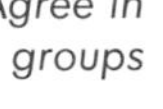
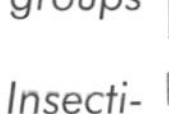

Terrestrial

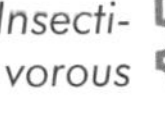

Agree in groups

Insecti-vorous

80–140 eggs

The paler coloration of its belly, and darker coloration on its back, which is greenish-black, helps to separate this species of fire-bellied toad from its relatives. In addition, it has fewer wart-like swellings over its back, clearly distinguishing it from the yellow-bellied toad.

EUROPEAN FIRE-BELLIED TOAD
The tips of the toes in this species show no bright coloration.

REPRODUCTION The European fire-bellied toad can be reluctant to spawn. Hibernation seems to be essential, with a temperature of 39°F (4°C) maintained for at least 8 weeks in late winter. A dense covering of weed and also heating the water to at least 77°F (25°C) may be necessary to induce breeding. In contrast to many frogs and toads, *Bombina* species mate with the male holding the female by her rear rather than front legs. Repeated egglaying may occur through the summer, the spawn usually being deposited in the waterweed. The young tadpoles grow fast, leaving the water as young frogs sometimes within 6 weeks.

YELLOW-BELLIED TOAD

Bombina variegata

2 in.
(5 cm)

Semi-aquatic

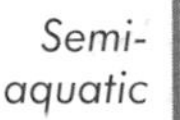

Terrestrial

Agree in groups

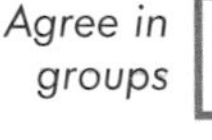

Insecti-vorous

25–150 eggs

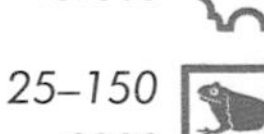

The belly coloration of these toads tends to be duller than that of the fire-bellied toads, although this depends to some extent on their respective diets, since the color is influenced by pigments in the toad's food. Yellow-bellied toads are found in southern Europe. There are four distinct subspecies distinguished by their size and coloration.

Their care is straightforward, and they can live for as long as 20 years. The vivarium should include an area of land as well as a stretch of water for swimming and breeding. When the water has to be changed, use a siphon to remove the dirty water so that the dry area is not disturbed.

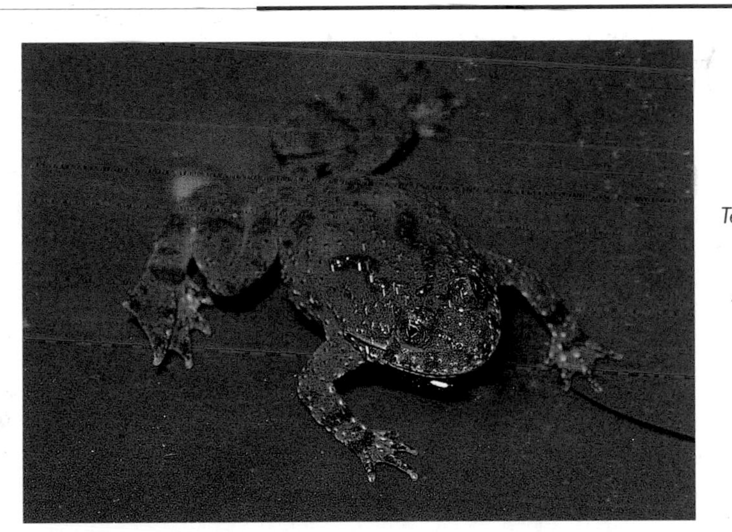

REPRODUCTION This is virtually identical to that of other *Bombina* species, with a cool overwintering period followed by warm water to stimulate breeding and the development of the tadpoles.

YELLOW-BELLIED TOAD
The "warts" on the back are very prominent in this species.

8 in. (20 cm)

Semi-aquatic

Terrestrial

Aggressive

Carnivorous

12,000–25,000 eggs

AMERICAN BULLFROG

Rana catesbiana

It is unusual to find these large frogs for sale as adults. Instead, tadpoles are often to be seen at aquatic stores, and even these are quite big. A red-eyed white strain is also available. Unfortunately, the larvae are highly predatory by nature, and if introduced into a pond they are likely to eat many of the residents. They develop into very large frogs which themselves are determined hunters. Under no circumstances should the tadpoles be introduced into a garden pond unless its perimeter is secure, for otherwise they could escape and devastate local wildlife, once their metamorphosis is complete.

A pond is undoubtedly the best way of keeping these frogs, in view of their size and nervous nature, which means they will not settle well in the confines of a vivarium. Snout injuries are not uncommon under these conditions, particularly if there is a shortage of retreats. An outdoor pond can be surrounded with bricks, with a wood-framed mesh anchored on the roof to prevent any escapes. The bullfrogs will spend much of their time in the water, although their large size and powerful limbs make it difficult to establish plants. They will take a wide variety of prey, with larger individuals eating pinkies without difficulty.

REPRODUCTION This is uncommon unless the frogs are housed in a pond. Mating takes place after hibernation, once the water temperature has risen above 68°F (20°C). Hatching is slow, sometimes taking more than 10 days, and although the tadpoles feed greedily, they will take 20 months to grow into frogs.

AMERICAN BULLFROG
Male bullfrogs have yellow throats and larger eardrums (the circular area behind each eye) than females.

AFRICAN BULLFROG

Pyxicephalus adspersus

This is another highly predatory species, which needs to be kept in a relatively warm environment with a soft substrate and moss. It spends much of its time partly buried here, its powerful hindlimbs equipped with shovel-like projections used to dig itself in. A large, deep water bowl should be supplied, with its contents changed regularly.

This species is also sometimes described as the Pac-man frog, because of its appearance. As it matures, folds of skin develop in the sides of its body. It can be even more aggressive than its American relative, and it is unwise to offer food by hand, because you may be bitten. Its mouth contains three sharp, toothlike swellings which can inflict a painful injury.

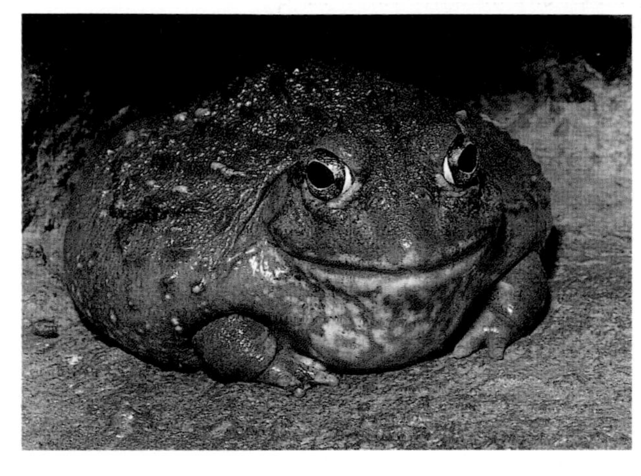

REPRODUCTION It is not safe to keep these bullfrogs together even for mating, because of the risk of cannibalism. They must be fed to saturation point beforehand, as a precaution. Males are larger, with yellow throats. They may attempt to swallow potential partners. A large greenhouse offers the best chance of successful spawning, allowing either partner to escape aggression.

AFRICAN BULLFROG
Keeping these frogs in dry surroundings, and then flooding their quarters, can stimulate breeding.

8 in. (20 cm)

Savannah

Terrestrial

Aggressive

Carnivorous

10,000–12,000 eggs

ASIATIC HORNED TOAD

Megophrys nasuta

The projections extending above the eyes of these toads give them a bizarre appearance. They help to provide camouflage, making the toad look like a leaf on the forest floor. Horned toads need warm, humid surroundings and a good covering of leaves and leaf litter on the floor of their quarters. Cork bark can be used to make retreats. A bowl of water should be placed in the vivarium. They feed readily on a variety of invertebrates.

REPRODUCTION It can be difficult to obtain pairs of these toads, the larger females being relatively scarce. Reducing both temperature and humidity in the vivarium for 2–3 weeks once a pair is well established may serve to trigger breeding. The female will spawn on wood overhanging water, with the tadpoles dropping into the water after hatching. After several days, they will feed at the water surface on flaked foods and should start to grow rapidly, sometimes leaving the water just 25 days after hatching. They need to be reared individually, because of their cannibalistic tendencies.

ASIATIC HORNED TOAD
Hidden on the forest floor, these toads grab prey as it passes within reach.

6 in. (15 cm)

Tropical woodland

Terrestrial

May eat smaller companions

Insectivorous

500–700 eggs

HELLBENDER

20 in. (50 cm)

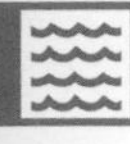
Aquatic

Carnivorous

May eat smaller companions

600 eggs

Cryptobranchus alleganiensis

Hellbenders are found in eastern North America. They live in fast-flowing stretches of water, to which they are well adapted. They lurk under rocks and in cavities on the bottom, and are mainly nocturnal. A large unheated aquarium, or better an indoor pond, is needed and the water must be well aerated. A small waterfall above a pond will help. Hellbenders normally feed on aquatic invertebrates, but they will also eat some prepared fish foods or even raw meat.

REPRODUCTION A cool period in the winter will condition hellbenders to breed later in the summer. The smaller male digs a hole where the female lays her eggs. He fertilizes them and buries them in the hole so they will not be washed away. The larval hellbenders emerge 8–12 weeks later.

HELLBENDER
The head of the hellbender is flattened, and its eyes are small. The hellbender is a protected species throughout its range.

MUDPUPPY

14 in. (35 cm)

Aquatic

Agree in groups

Carnivorous

150 eggs

Necturus maculosus

These primitive salamanders from southern Canada and northern USA also need to be kept in unheated water, which should be no warmer than 61°F (16°C) so that it can contain plenty of dissolved oxygen. Mudpuppies rely on their feathery gills to extract oxygen, like larvae, rather than developing into adult salamanders with lungs. They require similar conditions to the hellbender, with either sand or fine gravel on the base of the tank. They will eat aquatic invertebrates, including snails, as well as fish.

REPRODUCTION Females can be recognized as they swell in size with eggs. Mating takes place in fall, with the female retaining the packets of sperm, called spermatophores, in her body through the winter, before finally laying her eggs in the spring. These are concealed in a small den built by the female. She remains with her brood until they hatch around 2 months later, by which stage they are just under 1 in. (2.5 cm) long.

MUDPUPPY
The name of these salamanders comes from the rather squeaky sound of their call, said to resemble the bark of a puppy.

FIRE SALAMANDER

Salamandra salamandra

FIRE SALAMANDER *Overwinter fire salamanders at a temperature of about 45°F (7°C).*

The bright coloration serves as a warning: a poisonous substance is released from skin glands to deter predators. The appearance of these salamanders differs widely through their range, with at least 13 different forms being recognized. They are found in a wide area from central and southern Europe to Asia and North Africa. Some may have spots or larger blotches rather than lines of yellow running down their bodies. The appearance of orange, rather than yellow, fire salamanders does not appear to be a geographical phenomenon, as young of this color can come from a litter of yellow offspring. There is a wide variety in size and tail length among the subspecies. If possible, buy stock from one source at the same time, so you can be reasonably sure of starting out with salamanders of the same type.

Fire salamanders are easy to keep where there are plenty of retreats made of pieces of wood and bark. A water bowl should also be included, although these salamanders are mainly terrestrial. Nocturnal in their habits, they prefer to hunt for their invertebrate prey after dark.

REPRODUCTION Most fire salamanders produce larvae, but there are some populations which give birth to fully developed young. Mating takes place on land, with the female normally returning to water to give birth to her larvae. Small worms, such as whiteworms, and aphids make suitable rearing foods for them. Their gills will start to disappear between 3 and 5 months later. The water level should be lowered at this stage, and rocks provided, so the young salamanders can rest out of the water as their lungs start to function.

10 in. (25 cm)
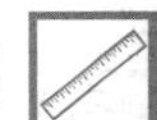

Temperate woodland

Terrestrial

Agree in groups

Insectivorous

25–75 eggs

5 in. (12 cm)

Temperate woodland

Terrestrial

Agree in groups

Carnivorous

8–16 eggs

RED-BACKED SALAMANDER

Plethodon cinereus

The plethodontid family of salamanders originate from southern Canada and eastern USA. They are often described as lungless salamanders, because they breathe through their skins, rather than using lungs. They are a primitive group and need cool, moist, well-ventilated surroundings, with clean dechlorinated water, if they are to thrive. Their secretive behavior helps them to avoid predators and also avoids desiccation. The red-backed salamander is a typical member of this group, hiding during the day and emerging to hunt invertebrates in the cool of the evening.

REPRODUCTION **Another feature of these salamanders is the care given to her brood by the female, who remains curled around her eggs. Her skin contains protective secretions which inhibit the growth of fungus on the eggs, which would impair their hatchability. The young emerge as miniature adults. Hatching takes up to 10 weeks.**

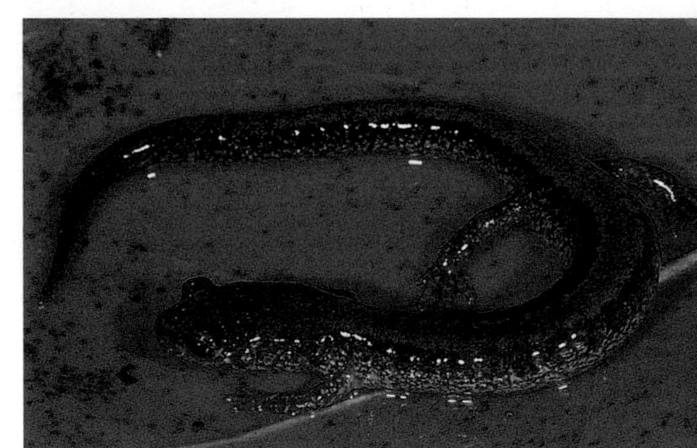

RED-BACKED SALAMANDER
These salamanders can climb very effectively, so keep their quarters covered.

8 in. (20 cm)

Temperate woodland

Terrestrial

Agree in groups

Carnivorous

30–45 eggs

SLIMY SALAMANDER

Plethodon glutinosus

Recent studies have revealed that what was previously thought to be just a single species in fact comprises at least 13 different species, which look remarkably similar but are genetically different. Slimy salamanders come from Canada and eastern USA. They are secretive by nature, hiding for long periods under rocks and similar retreats, which must be provided in their vivarium. They need relatively moist surroundings and should be allowed to hibernate in the winter.

REPRODUCTION **Males can be distinguished by the presence of a prominent gland under the lower jaw, and their cloacal region is more prominent than in females. The female lays in a damp concealed place, and stays with her eggs until they hatch. The young measure about 1 in. (2.5 cm) in length when they hatch 2–3 months later.**

SLIMY SALAMANDER
These salamanders produce a sticky body secretion as a defensive mechanism. Avoid handling them directly if possible.

RED SALAMANDER

Pseudotriton ruber

This species from eastern USA ranks among the most colorful of all salamanders. It needs a very damp and humid environment, but the ventilation in the vivarium also needs to be adequate to prevent the development of any harmful molds. Moss, wood, and slate can all be used to create an attractive yet functional arrangement. Provide dechlorinated water in a separate bowl, which can be easily changed several times a week. Red salamanders are often found near springs, where there is a continuous flow of water. Their environment should be cool, being kept at 41–50°F (5–10°C) in winter.

REPRODUCTION **The shape of the cloacal region allows these amphibians to be sexed quite easily; it is more prominent in males. Red salamanders need water for breeding and may feed in the water at this stage. The eggs must be kept well oxygenated, hatching up to 6 weeks after laying. Metamorphosis is very slow and may take 2 years.**

RED SALAMANDER
These colorful salamanders may live for nearly 20 years.

6 in. (15 cm)

Semi-aquatic

Terrestrial

Agree in groups

Carnivorous

100–150 eggs

LONG-TAILED SALAMANDER

Eurycea longicauda

Members of this genus from eastern Canada and USA vary quite widely in their habits, some being more aquatic than others. The long-tailed salamander spends some time on land, although it rarely strays far from water. Coloration varies, with a dark stripe running down the center of the back indicating a member of the southern subspecies, sometimes known as the three-lined salamander (*E.l. guttolineata*).

Leaf litter and moss can be used to create a substrate in the vivarium, which should also have water with broad-leafed aquatic plants. These salamanders must be kept cool, below about 61°F (16°C), so it may be necessary to transfer their vivarium outside to a shady spot when the weather is warm.

REPRODUCTION **Eggs are laid on leaves of aquatic plants, the larvae hatching 6 to 8 weeks later. They may remain in their aquatic phase for 7 months before emerging onto land. They are mature by 2 years old.**

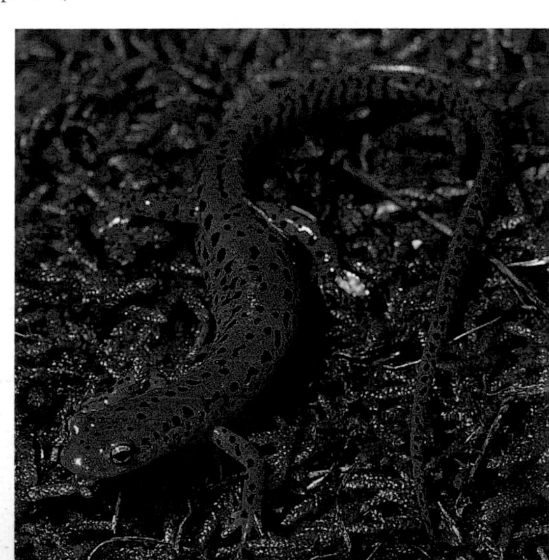

LONG-TAILED SALAMANDER
The tail of the long-tailed salamander is longer than the body.

6.5 in. (16 cm)

Temperate woodland

Terrestrial

Agree in groups

Carnivorous

8–25 eggs

MOUNTAIN DUSKY SALAMANDER

5 in. (12 cm)

Temperate woodland

Terrestrial

Agree in groups

Carnivorous

15–60 eggs

Desmognathus ochrophaeus

This salamander comes from south eastern Canada and eastern USA. It lives close to flowing water, such as small streams. Like other members of its family, it needs well oxygenated, clean water. There is considerable dispute over the taxonomy of these salamanders, which appear to be naturally variable in color. Individuals may vary from light brown through shades of gray to black, with spots or larger blotches of color which are equally variable. Purchase a small group from one source, so that you can be fairly sure they originated from the same locality.

REPRODUCTION Males are often identified by their larger head, with glandular swellings under the chin. To stimulate breeding, reduce the temperature in late winter. The female lays her eggs near water from late summer to early fall. She broods them, and keeps them moist until they hatch. This may take around 2 months, and in some cases up to 7 months. The young are about 1 in. (2.5 cm) long and more colorful than the adults.

MOUNTAIN DUSKY SALAMANDER
The maternal devotion of these salamanders is remarkable.

FOUR-TOED SALAMANDER

4 in. (10 cm)

Temperate woodland

Terrestrial

Agree in groups

Carnivorous

30 eggs

Hemidactylium scutatum

This unique species from North America can be identified without difficulty by the presence of just four toes on its hind feet. It is similar in its needs to the *Plethodon* species, needing a moist area of land and an area of water. This can be achieved if necessary by having a partition in the tank, stuck in place with silicone sealant. Use only ones recommended for use with fish tanks and sold through the larger aquatic outlets – other general building sealants may have toxic components in them, and amphibians are very susceptible to poisoning.

REPRODUCTION Mating takes place in the spring after winter dormancy. The eggs are often laid in the water, with the female remaining in attendance until the larvae emerge. This can take 2 months. Small aquatic live foods such as *Daphnia* can be given in moderation, although it is equally important not to pollute the water with uneaten food.

FOUR-TOED SALAMANDER
Reddish-brown upperparts are a feature of the four-toed salamander.

LONG-TOED SALAMANDER

Ambystoma macrodactylum

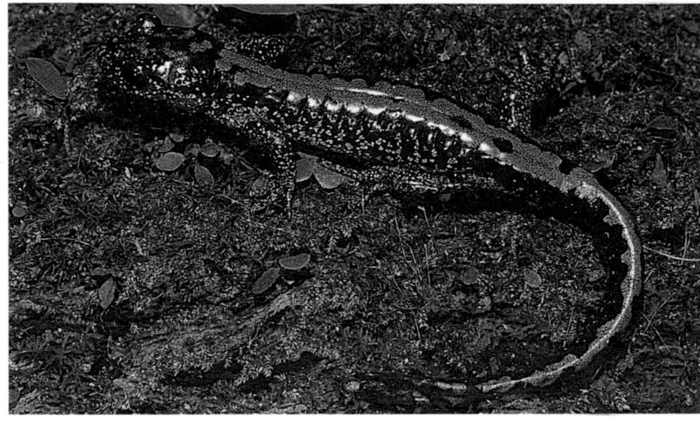

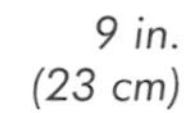

9 in. (23 cm)

Temperate woodland

Terrestrial

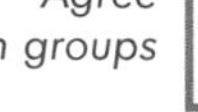

Agree in groups

Carnivorous

150 eggs

The most obvious feature of this salamander, in spite of its name, is the colorful band, varying from green to gold, running down the length of its back. In some cases this may be broken into blotches or even spots. The long-toed salamander comes from North America and tends to be more aquatic than some other *Ambystoma* species, and must have an area of water in its enclosure at least 6 in. (15 cm) deep. Adult long-toed salamanders will feed in the water and on land.

REPRODUCTION It will take up to 6 weeks for the eggs to hatch into larvae. Small aquatic creatures, especially those which can be safely cultured at home, such as brine shrimp, mosquito larvae, and whiteworm, can all be recommended as foods at this stage. It may take up to a year before the larvae complete their development into young salamanders, by which stage they may have grown to 4 in. (10 cm) long.

LONG-TOED SALAMANDER
The long-toed salamander is an egg-laying species.

TIGER SALAMANDER

Ambystoma tigrinum

13 in. (33 cm)

Temperate woodland

Terrestrial

Agree in groups

Carnivorous

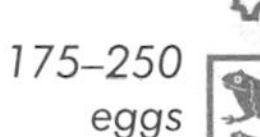

175–250 eggs

As many as 12 subspecies of the tiger salamander are recognized, and even then there may be considerable variation in appearance even between related individuals. They come from a wide area from southern Canada to Mexico. Dark brown or black markings are combined with olive-yellow coloration, which may predominate in some cases, notably in the case of the barred tiger salamander (*A.t. mavortium*).

Tiger salamanders can become quite tame, being less shy than some of their smaller relatives. They also have bigger appetites, eating small mice without difficulty. It is possible to keep them outside in a suitable secure enclosure, with retreats available and a large area of water, although they should be brought indoors for the winter in areas where the temperature dips to freezing point.

REPRODUCTION An enlarged cloacal area distinguishes males. Breeding is likely to be successful in an outdoor enclosure in early spring, after the salamanders have been kept cool over winter. The water should be at least 24 in. (60 cm) deep, with plenty of aquatic plants for spawning. Avoid heating the water to speed up hatching. The resulting larvae are less likely to thrive than those reared normally. Cannibalism may be a problem, and the young should be kept in small groups of similar size to reduce the risk.

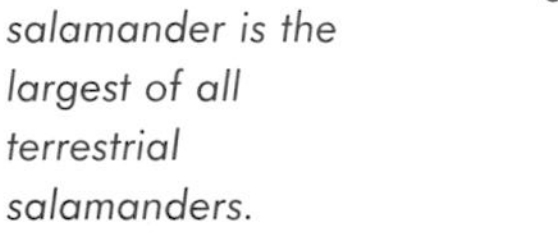

TIGER SALAMANDER
The tiger salamander is the largest of all terrestrial salamanders.

MARBLED SALAMANDER

Ambystoma opacum

The distinctive silvery gray and black markings of the marbled salamander make this one of the most attractive species. It comes from eastern USA and prefers drier conditions than other salamanders of the genus *Ambystoma*. It will thrive in an outdoor enclosure during summer. In view of its size, smaller prey should be provided than for the tiger salamander; a variety of invertebrates is acceptable, including worms and slugs.

4.5 in. (11 cm)

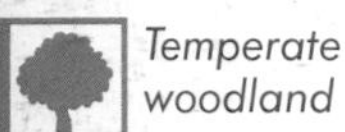

Temperate woodland

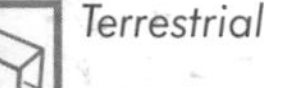

Terrestrial

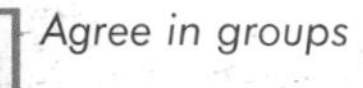

Agree in groups

Carnivorous

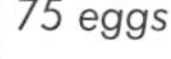

75 eggs

MARBLED SALAMANDER
These salamanders normally reach maturity by the time they are about 15 months old.

REPRODUCTION Male marbled salamanders are invariably more brightly marked than females, showing whiter banding in most cases, and marked swelling around the cloaca. The breeding habits of this species are different from others of the genus: mating takes place in the fall, with the female picking up the males spermatophore on land and subsequently laying her eggs on the floor of a dried-up temporary pool. She stays with them until the rain comes, and usually the larvae hatch in the water. Even in arid surroundings, however, the development of the young salamanders continues in the eggs, and they finally emerge in the spring as miniature adults about 3 in. (7 cm) long. The process takes about 5 months in total, whether or not an aquatic larval phase is involved. Clearly, it is much easier to replicate the necessary conditions outdoors rather than in a vivarium. It is necessary to make sure that the young salamanders will not be able to escape after completing the metamorphosis.

SPOTTED SALAMANDER

Ambystoma maculatum

As their name suggests, these salamanders have an attractive spotted appearance, some being more vividly marked than others. Their background color is a grayish-black, with yellow or reddish-orange spots arranged in two rows running from head to tail. Some spotted salamanders may have both these colors. This species from southern Canada and eastern USA will thrive in outdoor conditions, like those for the tiger salamander, but preferably more moist. Here there will be less risk of spotted salamanders developing fungal ailments, to which they can be susceptible, as they will be able to roam over a larger area. Damp spots, with logs and other retreats are essential, but part of the enclosure may be kept drier.

SPOTTED SALAMANDER
In spite of its striking appearance, this is a relatively shy species which often remains hidden for long periods.

REPRODUCTION Breeding can be achieved in spring, following hibernation. The salamanders will gather in a pool of water, where mating takes place. The eggs take 1–2 months to hatch, and it may be another 4 months before the young move onto land, when they will have grown to at least 2 in. (5 cm) long.

8 in. (20 cm)

Temperate woodland

Terrestrial

Agree in groups

Carnivorous

20–75 eggs

Will chlorine and similar chemicals added to our water harm amphibians?

Not necessarily, but it may be better to add a special dechlorinator to water provided in their quarters, especially in the case of highly aquatic species. Products of this type are sold for use in aquaria. Check that the product chosen is also effective against chloramine, which is now being widely used and could potentially be more harmful than chlorine itself!

AXOLOTL

Ambystoma mexicanum

AXOLOTL
The albino form of the axolotl is not found in the wild.

These strange amphibians are one of the best-known examples of neoteny: they remain in their immature larval stage, as large tadpoles, in which state they can even breed. But if the water level in an axolotl's quarters is gradually lowered over several months, this will cause it to start developing into a salamander, and its gills will shrink in size. Should the water level then be slowly restored, the axolotl will resume its aquatic lifestyle, with its gills becoming larger again. Axolotls also have other remarkable regenerative powers, such as the ability to regrow limbs which are sometimes lost in fights with others of their kind. They also seem relatively resistant to fungal ailments following injuries.

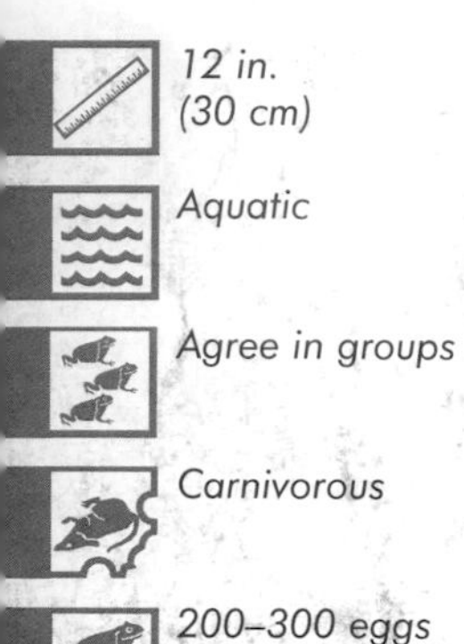

12 in. (30 cm)

Aquatic

Agree in groups

Carnivorous

200–300 eggs

Adding iodine to the water in its tank can also trigger metamorphosis in an axolotl. This trace element is taken in by the thyroid gland in the neck, altering the amphibian's rate of metabolism. Other amphibian tadpoles which are unusually slow to transform may also be deficient in iodine.

Axolotls have been kept and studied around the world because of their strange physiology, although their native habitat is a small area of Mexico, Lakes Xochimilcho and Chalco. They are considered to be under threat in the wild, but huge numbers are bred in many countries, so obtaining stock is not difficult.

Two main color variants are available, although only the dark sooty brown form occurs in the wild. The albino variant emerged many years ago, as domestication was proceeding, and today it is probably just as common as the natural form. It shows some red markings, particularly on its gills, which are in fact blood vessels. Pied variants, with black and white coloration, plus other colors such as olive, and even a golden shade have been recorded.

A reasonably spacious aquarium will be needed, although these are not particularly active amphibians. No heating indoors is necessary, but their water must be kept clean, and the use of a water conditioner (as sold for fish) is recommended. The water depth should be about 12 in. (30 cm). Feeding is straightforward, as axolotls will eat many invertebrates. They will also take inert foods, such as pieces of meat or even foodsticks, if these are moved around close to their faces.

REPRODUCTION Breeding takes place in spring, with the female depositing her eggs on aquatic plants. The young hatch about 2 weeks later. At this stage they measure about ½ in. (1.5 cm) and are miniature adults. Having digested their yolk sacs, they start feeding. Powdered fish food can be used at this stage, with other foods such as whiteworm being introduced as the young grow larger. They can breed at a year old.

AXOLOTL
Axolotls can live for over 20 years.

EUROPEAN COMMON NEWT

Triturus vulgaris

4 in. (11 cm)

Semi-aquatic

Terrestrial

Agree in groups

Carnivorous

200–300 eggs

These newts spend part of their time on land, seeking out moist areas where they can hide, returning to the water for spawning in the spring. Sometimes they will spend much of the summer in the water as well. Their aquatic nature is shown by the webbing on their hind feet. In a sunny locality these newts can be kept satisfactorily in a secure outdoor enclosure with an area of water containing a dense covering of oxygenating plants in which they can breed. They will feed on a variety of invertebrates, both in and out of water. They will also prey on other amphibians, often eating large numbers of frog tadpoles in ponds.

EUROPEAN COMMON NEWT *Females may lay as many as 300 eggs.*

REPRODUCTION Before breeding, males develop crests along the back to the tail. For the rest of the year they can be re-cognized by their dark spotting. Mating occurs in spring when both sexes return to the water. The female lays her eggs individually on the leaves of aquatic plants over several weeks. She disguises them using her hind legs to fold the leaves over, so the eggs are hidden. Hatching occurs within 2 weeks or so, and the young measuring about 1 in. (2.5 cm), come onto land in 2–3 months, depending on the temperature. They will then roam on land until they reach breeding age in 2 years.

ALPINE NEWT

Triturus alpestris

5 in. (13 cm)

Semi-aquatic

Terrestrial

Agree in groups

Carnivorous

300–400 eggs

Often found at relatively high altitudes, some fairly isolated populations of these newts are established in countries from Spain to Greece. As might be expected, alpine newts are remarkably hardy and often highly aquatic by nature, depending on the subspecies. The Italian subspecies (*T.a. apuanus*) lives almost entirely in water. These newts can be easily maintained outdoors, but they must be prevented from escaping. Newts climb very well, so their quarters, whether indoors or out, must be securely covered.

ALPINE NEWT *Alpine newts become an unusual blue color at breeding time.*

REPRODUCTION Males are particularly attractive when in breeding condition. They have a dark-blue back with lighter white markings, and orange underparts. Blue is a rare color in amphibians. Females of this species are much duller, being predominantly brownish. Their eggs take up to a month to hatch, and metamorphosis is complete within about 3 months, by which time the young newts will be over 1 in. (2.5 cm) long.

NORTHERN CRESTED NEWT

Triturus cristatus

These relatively large newts from northern Europe can be easily identified, both by their size and by their blackish bodies with orange underparts. The spectacular appearance of this newt has led to fears for its numbers in the wild, and it is protected in several countries through its wide distribution. Captive-bred stock is available, however, and these newts will thrive especially if given access to a pond, rather than being housed in an ordinary vivarium. Breeding is also more likely in such surroundings. There is a southern form of crested newt which used to be considered a subspecies, but is now accorded specific status in its own right as *Triturus carnifex*. This also applies in the case of the Caucasian crested newt (*T. karelini*) and the Danube species (*T. dobrogicus*).

6.5 in. (16 cm)

Semi-aquatic

Terrestrial

Agree in groups

Carnivorous

200–250 eggs

NORTHERN CRESTED NEWT
White spots on the sides of the body help to identify this species.

REPRODUCTION At the outset of the breeding period, following a period of hibernation, male crested newts start to develop a tall dorsal crest. They prefer to breed in deep water, and broad-leafed plants such as watercress are often favored as nesting sites. In other aspects they are similar to other members of the genus, with young crested newts being mature when approximately 2 years old.

RED-SPOTTED NEWT

Notophthalmus viridescens

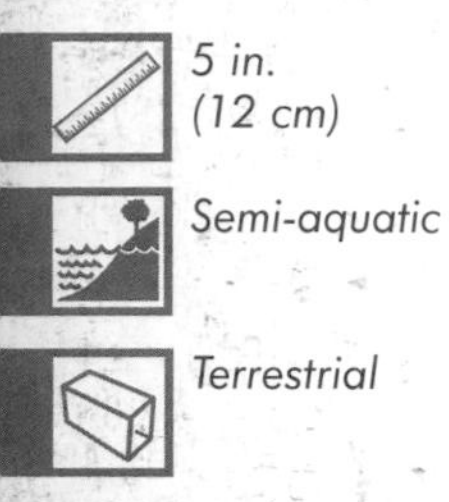

5 in. (12 cm)

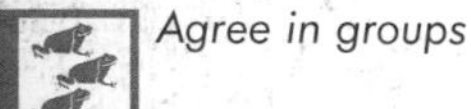

Semi-aquatic

Terrestrial

Agree in groups

Carnivorous

300–350 eggs

This species comes from North America and in spite of their name, not all subspecies of the red-spotted newts actually have red spots, which are highlighted with black borders, on their backs. In the broken-striped newt (*N.v. dorsalis*), for example, the spots are generally elongated into stripes, but the Florida peninsula newt *(N.v. piaropicola)* usually has just a dark-brown back. The red markings, as is often the case with amphibians, are a sign of the presence of an unpleasant toxin in their skin, and if they need to be handled, disposable plastic gloves should be worn.

Adult red-spotted newts are mainly aquatic and prefer to hide among water plants. The water must be kept clean, because they are very susceptible to fungal disease, which can be spread through the water. They prefer a temperature of

RED-SPOTTED NEWT
The red eft has a stunning coloration, but regrettably it is only temporary.

REPRODUCTION There is a distinct division in the lives of red-spotted newts, depending on their age. Having metamorphosed at about 3 months old, the young newts are very different in appearance from adults. Their skins give them more protection against desiccation, and they are bright red in color, which has led to their being known as red efts. The red spots of adult newts may already be discernible. A vivarium for red efts should consist predominantly of land, with just a water bowl being provided. They will eat terrestrial invertebrates. As they become mature, around the age of 2–4 years, the red efts return to water, changing into recognizable red-spotted newts.

RED-SPOTTED NEWT
Red-spotted newts need to be kept relatively warm.

MARBLED NEWT

Triturus marmoratus

The marbled newt is found in the Iberian peninsula and France. It is widely kept and bred, so obtaining stock should not be difficult. It tends to be less aquatic in its habits than related species, and its quarters should be designed accordingly. Sphagnum moss should be used as the substrate, with hiding places formed of cork bark and stones. Water should be provided in a dish large enough to allow the newts to immerse themselves completely. Invertebrates should form the basis of their diet.

REPRODUCTION The orange stripe down the center of the back is characteristic of females. Males develop a crest when they enter the water for breeding. Hibernation over winter is necessary before spawning in spring. Adult newts in good condition should be kept in an environment just above freezing, in a container of sphagnum moss. Some breeders use a refrigerator, because the temperature can be accurately controlled. Females return to the water first. A large area should be provided for spawning. Newts will often lay eggs on corrugated plastic sheeting rather than aquatic plants.

MARBLED NEWT
This is one of the most attractive of all the Triturus species.

7 in. (17 cm)

Semi-aquatic

Terrestrial

Agree in groups

Carnivorous

150–250 eggs

7 in. (18 cm)

Semi-aquatic

Terrestrial

Agree in groups

Carnivorous

100–150 eggs

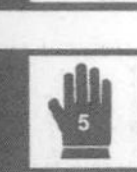

ROUGH-SKINNED NEWT

Taricha granulosa

These largish newts from North America must be handled carefully, because of a highly toxic skin secretion – you should wear plastic gloves. The skin is rough, with projections on it, and the underparts vary in color from yellow to orange.

A damp vivarium with a water bowl will suffice until the breeding season, when they will need a large area of water.

REPRODUCTION It is difficult to identify the sexes with certainty outside the breeding period, but then males can be recognized by the swollen area surrounding their cloacal orifice. A period of hibernation is necessary. Water for spawning should be about 6 in. (15 cm) deep, and Canadian pondweed should be provided for the female to conceal her eggs. She lays each one with great care, tucking the leaf over afterward to conceal its presence. Hatching can take up to 10 weeks, with young newts remaining for as long as a year in their larval phase. But they usually turn into adults by 4–6 months.

ROUGH-SKINNED NEWT

These newts can become quite tame and learn to take food off blunt-ended forceps.

I'm breeding large numbers of a native amphibian. *Seek advice from a local wildlife group first. There are a number of factors which have to be considered if schemes of this type are to work, and you may need to seek official approval. Nevertheless, if carried out successfully, such projects can be immensely rewarding. The deliberate release of non-native species is likely to be illegal.*

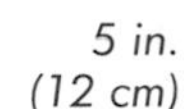

JAPANESE FIRE-BELLIED NEWT

Cynops pyrrhogaster

These large newts are easy to care for, and can be kept in mainly aquatic conditions throughout the year. Even so, they should have an area of rocks available, so they can leave the water for periods. The stones need to be carefully arranged so they cannot be dislodged, which might injure the newts. These newts can be kept outside in a pond through the warmer months, as long as there is no risk of escape. Japanese fire-bellied newts feed on aquatic invertebrates and worms.

REPRODUCTION Keeping these newts cool through winter will increase the chances of successful spawning the following spring. Eggs are laid individually by the female, and need to be transferred to a separate container for hatching, so the larvae are not at risk of being eaten. Hatching takes between 2 weeks and 1 month. Rearing is not a problem as long as a good food supply is maintained. The young should complete their metamorphosis in another 3 months.

JAPANESE FIRE-BELLIED NEWT
The fiery red underparts extend beyond the base of the tail. Some black blotches are also normally visible here.

5 in. (12 cm) 

Semi-aquatic 

Terrestrial 

Agree in groups 

Carnivorous 

150–300 eggs 

CROCODILE NEWT

Tylotriton verrucosus

This species originates from Indochina and western China. It is also known less commonly as the mandarin salamander, which is perhaps a more accurate reflection of its lifestyle, since it is not truly aquatic, venturing into water only for breeding. It is also known as the emperor newt. This species has only become available in recent years.

Crocodile newts inhabit areas of damp montane forest and need to be housed accordingly, in a vivarium lined with sphagnum moss and similar materials, with a bowl of water. The temperature should be kept at 64–68°F (18–20°C) for most of the year, with a slight drop in winter to encourage breeding in spring.

REPRODUCTION Move the newts in spring to more water. The female is considerably larger. She will lay her eggs on the leaves of plants. The eggs should hatch in about 4 days. It then takes 4–11 months for the larvae to complete the aquatic phase in their development, during which time they need to be kept fairly cool.

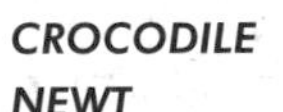

CROCODILE NEWT
Food for these newts can include small snails.

7 in. (18 cm)

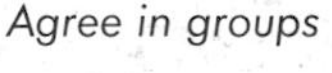

Semi-aquatic 

Terrestrial 

Agree in groups 

Carnivorous 

50 eggs

INVERTEBRATES

INSECTS

Insects are characterized by the presence of three pairs of legs, and their body is divided into three distinct segments – the head, thorax, and abdomen.

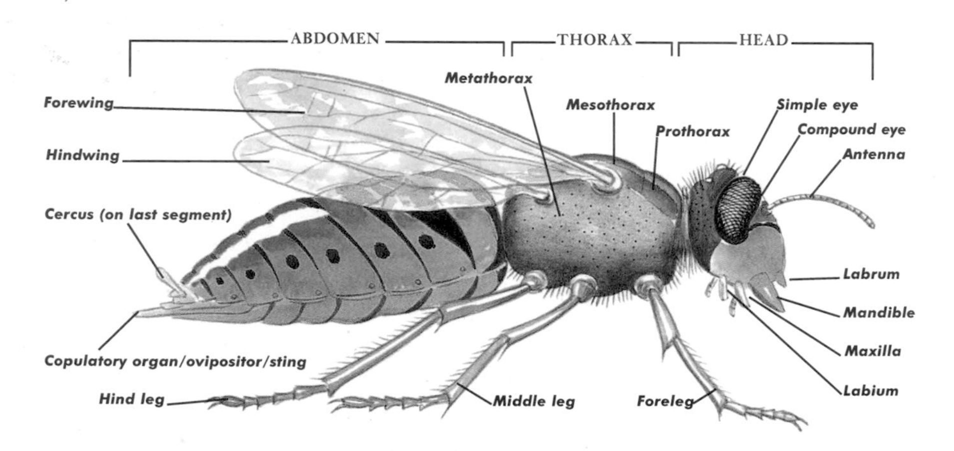

The typical appearance of an insect.

About 800,000 different species of insect have already been identified, and millions more may yet await discovery, although some will almost inevitably have become extinct in recent years as a result of human activity without their existence even being known. The greatest diversity of insects are found in the world's rainforests.

In spite of their small size insects can be very vicious, as shown by the case of the praying mantis, where the male is likely to be eaten by the female after mating. Stick insects are the most widely kept family (ignoring bees, which are kept commercially).

A close-up of a praying mantis, showing its compound eyes and mouthparts. Camouflage is often used by insects, whether for predatory or protective purposes.

ARACHNIDS

Members of this group may sometimes be confused with insects, although they can be easily distinguished by the fact that they have eight rather than six legs. The group includes all spiders, including the so-called tarantulas, as well as scorpions, mites, and ticks.

In recent years, the keeping and breeding of tarantulas has become especially popular. This is because these spiders are fairly undemanding in their requirements and often spectacular in appearance. A specialist collection can be built up and maintained at relatively low cost, in spite of the fact that these arachnids need to be housed individually for virtually all their lives.

Not everyone finds spiders appealing – the term "arachnophobia" has been coined for those who have an intense fear of them. The arachnids also include mites and ticks.

Much still remains to be resolved about their taxonomy, and new species are still being discovered. Captive breeding is now becoming increasingly commonplace, however, to the extent that even unusual species can soon prove prolific if kept under the right conditions. Spiders' high reproductive potential, producing hundreds of spiderlings from a single batch of eggs, has helped to ensure the widespread availability of stock to enthusiasts.

Some spiders and all scorpions produce venoms, and tarantulas have irritating body hairs as well, so they are creatures to look at rather than to handle.

OTHER INVERTEBRATES

Various other invertebrates may occasionally be available, including a number of different millipedes. Unlike their centipede cousins, millipedes are vegetarian in their feeding habits, although they are equally secretive. Bright colors on their bodies are likely to be indicative of toxic substances in their skins, so that care needs to be taken when they are handled. Always wash your hands afterwards, because otherwise if you wipe your eye this could cause severe irritation.

These millipedes are obviously not pets in the sense that they can be handled regularly, although they are attractive to look at and their behavior may be interesting to observe. In fact, it is not generally a good idea to handle any invertebrates more than strictly necessary, even those that are not toxic, because of the risk of injuring them. Always remember that most invertebrates are surprisingly agile and may wriggle out of your grasp and fall to the floor. It is always safer to move them in a small container with a lid.

Handling some invertebrates calls for particular care. Special forceps are recommended when picking up scorpions, to avoid being stung.

2 in.
(6 cm)

Temperate woodland

Terrestrial

Agree in groups

Vegetarian

50 eggs

STAG BEETLE

Lucanus cervus

There are more than 900 different species of stag beetle, widely distributed around the world. They tend to be relatively large, but there are some small species measuring less than 1/2 in. (1 cm) in length. Those from temperate areas can be kept easily in a vivarium, decorated with a floor covering of leaves on top of soil, along with old wood and moss. They like to feed mainly on vegetable matter.

REPRODUCTION **Rotting wood is essential for breeding many of these species, because the female will lay her eggs there, and the resulting larvae will then burrow into the wood and consume it before ultimately emerging as adult beetles. This process may take several years. Male stag beetles can be recognized by their prominent pincers, not unlike a stag's antlers and also used for fighting with other males, which is how these beetles acquired their name. Amazingly, in spite of their size, they are unable to use these modified jaws to bite, because the associated muscles are weak. Instead, the battle is simply a wrestling match.**

STAG BEETLE
Stag beetles thrive in vivarium surroundings, provided that their quarters are not too dry.

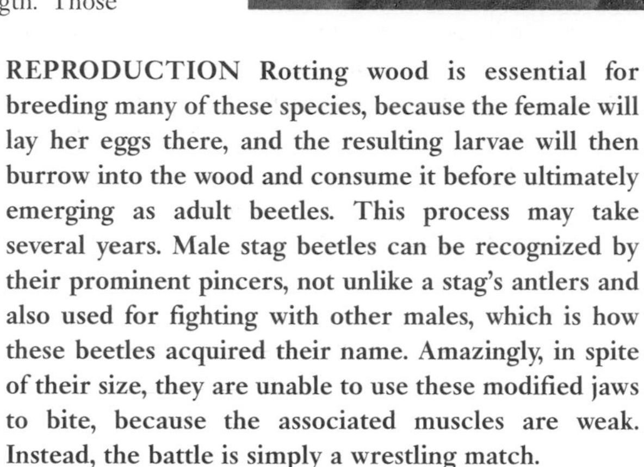

2 in.
(6 cm)

Temperate woodland

Arboreal

Agree in groups

Vegetarian

100–200 eggs

NORTH AMERICAN WALKING STICK

Diapheromera femorata

These cryptic members of the phasmid family resemble sticks in appearance, and live on plants. Their camouflage means that they can largely escape detection by predators such as birds or reptiles that would normally eat insects. Their legs enable them to grip very effectively, and to move along a branch or leaf when necessary. Walking sticks are usually known in Europe as stick insects.

REPRODUCTION **Eggs are scattered around their quarters, and resemble seeds. The hatching period is extended, presumably because if all the young walking sticks hatched at the same time, and were faced with a shortage of food in the wild, the entire population would suffer. They emerge from their eggs as miniature adults and grow through a series of molts.**

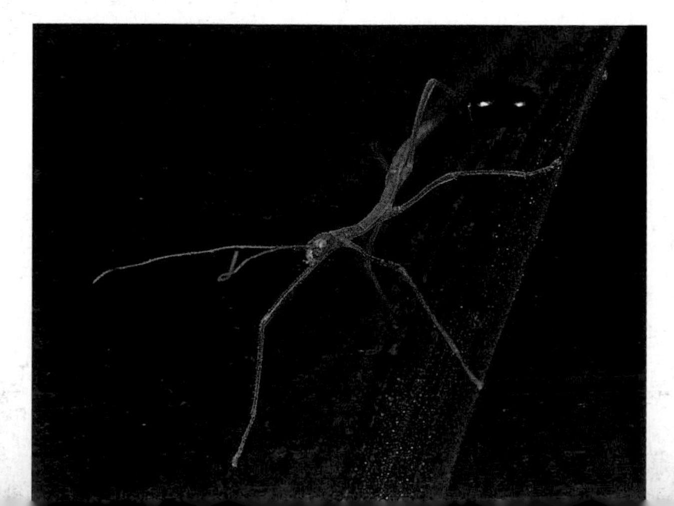

NORTH AMERICAN WALKING STICK
The North American walking stick is a popular pet in its homeland, but elsewhere other stick insects are more often kept. This hatchling has its empty egg still clinging to its leg.

INDIAN STICK INSECT

Carausius morosus

Few invertebrates are as easy to keep as these stick insects, recognizable by their thinnish green bodies with red streaks on their legs. They feed readily on bramble leaves, the staple food of this whole group of phasmids though they will also eat privet leaves. The stems of bramble should be cut and placed in a jar of water so they will not wilt, but choose a narrow-necked vessel, because otherwise the young could fall in and drown. Misting the leaves lightly will provide the insects with enough drinking water. Their quarters should be light and airy, so there is no risk of molds developing. A white paper lining on the floor of the enclosure makes it easy to clean up the insects' droppings and also allows you to spot their eggs easily.

If touched, stick insects show a remarkable ability to feign death. They simply fall to the ground, looking like a broken twig. After some time, they will stretch out and climb back onto a plant. They generally live for about a year.

REPRODUCTION There are virtually no male Indian stick insects. Instead, females reproduce without mating, with the young obtaining their genes entirely from their mother. Even when males do occur, they have not been seen to mate with females. They can be identified by their smaller size, and the red underside of their thorax.

Mature females produce a few eggs regularly over a long period. The eggs should be transferred to separate quarters and watched for the emergence of the nymphs. This will take place after 1–2 months, or later in some cases. They should not be allowed to become too dry. Avoid handling the nymphs directly because they are very delicate, and their legs are easily damaged at this stage. Use a paintbrush when you want to move them.

They will need more space as they grow, because if they are kept confined they eat each other's legs. This can be catastrophic for a stick insect, since it will no longer be able to climb properly and reach its food. It is also vital that their quarters are tall enough to allow them to hang vertically downward when molting, in order to free themselves properly from their old skin. Nymphs complete their development by about 5 months old and are likely to start laying eggs soon after.

4 in. (10 cm)

Tropical woodland

Arboreal

Agree well in groups

Vegetarian

100–200 eggs

INDIAN STICK INSECT

A regular supply of bramble or privet is needed for these stick insects.

PRAYING MANTIS

Mantis religiosa

The name of these mantids comes from the way in which they fold their front legs, while adopting a rather upright pose, not unlike a worshipper at prayer. There are more than 2,000 different species, with a fossil history dating back over 36 million years. Most praying mantids kept as pets come from the tropics and require a similar enclosure to walking stick insects, but they have an aggressive lifestyle, being predatory hunters. If an insect comes within reach, the mantis will strike with its front legs at a speed faster than the eye can follow, grabbing its prey in a fraction of a second.

4 in. (10 cm)

Tropical woodland

Arboreal

Aggressive

Insectivorous

20–300 eggs

REPRODUCTION This is a particularly hazardous time for the male, because he is likely to be decapitated by his mate. Males are much smaller and more agile than females, which helps them to survive. Feed the female well before mating, so she is less likely to be hungry. Finding the male's tiny empty sperm sac, about 1/10 in. (0.25 cm) long, on the floor of their quarters indicates that mating has probably taken place. The female lays her eggs in a special egg case or ootheca, and she may defend it from attack by other insects. The ootheca is best removed and incubated at a temperature of 84°F (29°C), with the young emerging about a month later. Initially, they can be reared together on a diet of fruit flies, but will soon need separating as they grow larger. A damp pad of absorbent cotton will provide moisture. They may mature in 2 weeks, with females laying up to 6 times during their life of about 3 months.

PRAYING MANTIS
Appearances can be deceptive. The praying mantis is a highly effective predator.

WHICH IS THE BEST INVERTEBRATE AS A CHILD'S PET? *Stick insects are very popular, in spite of the fact that they are not active by nature. They are easy to keep, though, and some species, such as the Indian (see page 141) can be picked up easily without any danger to the handler. However, never pull a stick insect off a branch. There are small hooks at the bottom of their feet which anchor them to twigs and pulling could result in loss of one or more legs.*

GIANT SPINY STICK INSECT

6 in.
(15 cm)

Tropical woodland

Terrestrial

Agree well in groups

Vegetarian

50–100 eggs

Eurycantha calarata

This stick insect was first found in New Guinea in 1978, and since then it has become quite widely distributed. Its habits are unusual: rather than climbing, it spends much of its time on the ground hiding under vegetation. It will consume a wide variety of greens, including grass. Handling needs to be carried out carefully, because these insects are well protected with a range of spines down the sides of their body. It is better to lift them gently from underneath.

REPRODUCTION Males have a much longer spine at the top of the first segment of their hind legs, and they are also about 1 in. (2.5 cm) shorter than females. Their eggs are cylindrical and can take up to 7 months to hatch, being kept in slightly moist sand to prevent them drying up. Young are often greenish in color, compared with the brownish-gray appearance of the adults.

GIANT SPINY STICK INSECT
These stick insects can defend themselves with their leg spines if threatened.

GIANT PRICKLY STICK INSECT

8 in.
(20 cm)

Tropical woodland

Arboreal

Agree well in groups

Vegetarian

100–1,000 eggs

Extatosoma tiaratum

This species is more widely kept than the giant spiny stick insect, although there can be confusion between them because of the similarity in their names. It is also sometimes called Macleay's specter. Today's captive stock traces its origins to a place in Queensland, Australia, where the first specimens were collected in the 1960s. Eucalyptus leaves form its normal diet, but it readily takes bramble. Roots of bramble can be grown quite easily in pots, soon producing shoots so that you can maintain a constant supply of food for the insects through the year.

REPRODUCTION Males have fully functional wings and are smaller than females, so that distinguishing the sexes is straightforward. Females can also reproduce without mating. The eggs must be kept in a reasonably warm environment, on moist sand. It can take between 6 and 9 months before they hatch, and possibly twice as long in the case of parthenogenetic eggs. The young are brownish in color, female nymphs being darker.

GIANT PRICKLY STICK INSECT
The development of these stick insects is complete by the time they are 5 months old.

JUNGLE NYMPH

Heteropteryx dilatata

Females of this large stick insect from Malaysia are truly stunning, but need very careful handling because of their long, sharp spines, especially those on the hind legs. They look more like leaves than sticks, in view of the width of their body.

Feeding is straightforward: bramble leaves are acceptable, also leaves of other plants such as rose and oak. Branches should be fairly sturdy, to support the weight of these insects. They need to be kept warm, with fairly high humidity and good ventilation. Spraying the leaves provides them with water.

JUNGLE NYMPH
Jungle nymphs do not live on the ground.

REPRODUCTION Sexing is easy, males are smaller and brown, while females are lime green. These insects are slow to mature, being unlikely to breed before a year old. The eggs are buried in a pot of sand in their quarters. It can be 16 months before the nymphs emerge, with the hatching rate tending to be low. The female's distinctive coloration starts to emerge after her third molt.

7 in. (18 cm)

Tropical woodland

Arboreal

Agree well in groups

Vegetarian

100 eggs

MADAGASCAN PINK-WINGED STICK INSECT

Sipyloidea sipylus

These slender stick insects are able to fly as a last resort if danger threatens, but they usually keep their wings hidden along their back. Twitching of the legs is an indicator that flight may be imminent. Try to avoid placing pieces of bramble with sharp thorns in their quarters, because they could tear their wings if they became snagged.

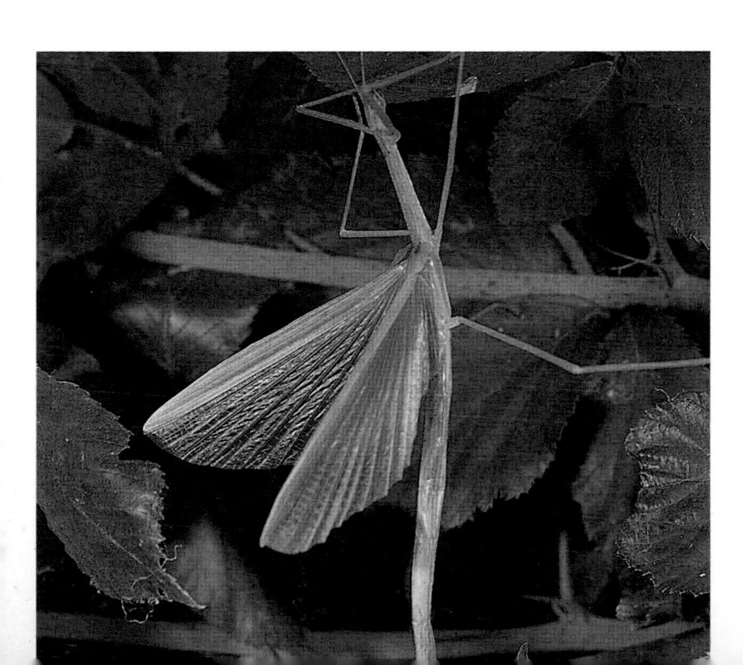

REPRODUCTION Males resemble females, but are smaller in size. This species often replicates parthenogenetically. The females glue the eggs around their quarters. It is not a good idea to shift the eggs, and they should be left to hatch, which will occur from 6 weeks on. The rather delicate nymphs will be green at this stage. They should be offered young shoots of bramble if possible, to encourage them to start eating. They often seem to eat more readily if kept with older individuals.

MADAGASCAN PINK-WINGED STICK INSECT
Pink-winged stick insects lay gray eggs.

3.5 in. (9 cm)

Tropical woodland

Arboreal
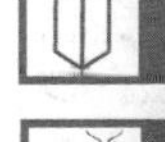

Agree well in groups

Vegetarian

100 eggs

MEXICAN RED-KNEED TARANTULA

Brachypelma smithi

In spite of their reputation, large spiders like the Mexican red-kneed are not generally dangerous, although they can give a nip rather like a bee sting if mishandled. The hairs on their body can also prove irritating, but since there is usually little need to handle tarantulas regularly, these points need not be of great concern. Note, though, that these spiders are surprisingly fragile. A fall is liable to rupture their abdomen, with fatal consequences.

The Mexican red-kneed is a burrowing species, and its vivarium should be designed accordingly. Broken flowerpots half buried in the substrate provide a very acceptable retreat. These spiders live in arid places and avoid dehydration by staying underground in their burrows during the day when the sun is at its hottest. Dew forms inside the entrance to the burrow at night, raising the humidity. These spiders will die quite rapidly from desiccation, and a relative humidity reading of 70–80 percent in their vivarium is necessary. This can be measured easily with a simple hygrometer, as sold in garden centers.

Prey can include crickets, sprinkled with a suitable vitamin and mineral supplement.

2.5 in. (6 cm)

Savannah

Terrestrial

Aggressive

Insectivorous

100–400 eggs

REPRODUCTION It is important to keep tarantulas on their own, because otherwise they may fight to the death. Even mating can be dangerous. Sexing of adults can be carried out without too much difficulty, because males have smaller abdomens. Closer examination should then reveal the presence of spurs on their front legs, coupled with swellings on each of the pedipalps near the mouth resembling miniature boxing gloves.

The spiders should be well fed before being introduced for breeding, to lessen the risk of aggression. The male displays, and if she responds positively he rushes across to mate with her as quickly as possible. He should then be transferred back to his quarters. The eggs are laid, often months later, in an egg sac, from which the young spiderlings should emerge within 4 months. They grow slowly and are mature by 5 years old.

MEXICAN RED-KNEED TARANTULA
Mexican red-knees are among the longest-living tarantulas. Females may live more than 30 years.

PINK-TOED TARANTULA

Avicularia avicularia

PINK-TOED TARANTULA
The black areas of this spider have an iridescent sheen.

1.5 in. (4 cm)

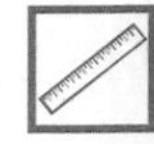

Tropical woodland

Arboreal

Aggressive

Insectivorous

100–400 eggs

These tarantulas are climbers, and will build a nest off the ground. Their basic coloration is blackish, with pink areas on their limbs bordering on orange in some cases. Pink-toed tarantulas come from northern South America and construct a web in the branches that serves rather like a hammock where the spider rests.

The spiders' quarters should include branches for climbing. Once settled, these tarantulas will tend to lose their shyness, particularly if they are left undisturbed. As might be expected, they prefer to feed on winged invertebrates, wax moths being ideal. Otherwise, crickets are a useful standby.

REPRODUCTION Males of this species are smaller in size. After mating, the female produces the egg sac within the security of her nest. When the young hatch, they are pinkish in color, with black areas extending upward from their toes, the reverse of the adult pattern.

CHILEAN ROSE TARANTULA

Grammostola cala

1.5 in. (4 cm)

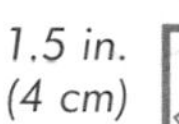

Tropical rainforest

Terrestrial

Aggressive

Insectivorous

100–400 eggs

This species is also sometimes known as the Chilean red-back, not to be confused with the deadly Australian red-back (*Latrodectus mactans*), which is not a close relative. Nor is it the same species as the Chilean pink tarantula (*G. spatulata*), which spends much of its time on the surface, rather than burrowing like the Chilean rose tarantula. A setup similar to that for the Mexican red-kneed (see page 146) suits this species well. There should be a fairly high level of humidity, and a shallow dish of water should be provided. These spiders will eat many kinds of small invertebrates.

REPRODUCTION The young take $1\frac{1}{2}$ to $2\frac{1}{2}$ months to hatch, after which they need to be reared separately. As with other tarantulas, males of this species have a relatively short lifespan.

CHILEAN ROSE TARANTULA
The Chilean rose tarantula is pale brown and pink in color.

4 in. (10 cm)

Tropical rainforest

Terrestrial

Aggressive

Insectivorous

100–400 eggs

GOLIATH BIRD EATER

Theraphosa leblondi

This huge species from northern South America has a leg span about as big as a dinner plate. It is relatively rare, and as a result is correspondingly expensive. In spite of its name, it does not eat birds, and spends most of its time on the floor of tropical forests. It prefers rather marshy areas, in which it burrows underground. During the rainy season these spiders are forced to spend more time above ground, since their usual retreats are then flooded.

In view of their size, these tarantulas need larger quarters than average. A dish of water is essential. They are not difficult to maintain on a diet of large invertebrates such as crickets, and can consume several of these at one sitting.

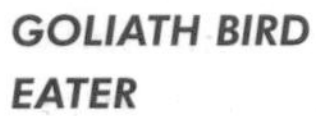

GOLIATH BIRD EATER

The chick in the picture gives a sense of scale, but of course these spiders do not actually eat birds.

REPRODUCTION Very little is known about the reproductive habits of this species. It is not yet being bred as often as other tarantulas.

2.5 in. (6 cm)

Desert

Terrestrial

Aggressive

Insectivorous

100–400 eggs

MEXICAN BLOND TARANTULA

Aphonopelma chalcodes

This spider's golden brown coloration has led to its also being known as the palomino, after the distinctive type of horse. Occurring in a similar type of terrain, it needs an identical setup to that recommended for the Mexican red-kneed tarantula (see page 146).

As it lives naturally in dry burrows, it is used to fairly low humidity. However, it may be a good idea to spray the vivarium gently with warm water when molting is imminent. Tarantulas lose their appetite before molting, the frequency of which depends in part on the age of the spider.

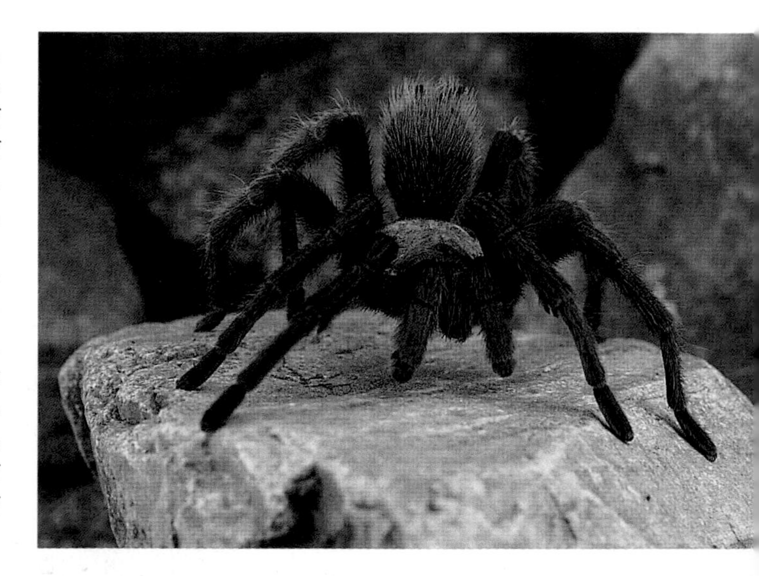

MEXICAN BLOND TARANTULA

The Mexican blond tarantula can be a fairly nervous species.

REPRODUCTION Captive breeding is not common, but it should not present any major problems once you have a true pair of spiders. They must be well fed before mating, so that the female will be less aggressive toward the male.

ZEBRA TARANTULA

Aphonopelma seemanni

This is one of the best-known tarantulas originating from Costa Rica. The creamy-white stripes on its black legs have given rise to its name. The Costa Rican zebra tarantula is a burrowing species and can move very fast. Plants are sometimes recommended to provide cover, otherwise these spiders can prove to be rather nervous, especially when first obtained. In the wild, their burrows may extend to a depth of about 5 in. (12.5 cm), so a good depth of substrate is recommended. The entrance is often carefully concealed with leaves, which should be provided.

REPRODUCTION Breeding has been achieved successfully in a vivarium, but it can sometimes be difficult to persuade these spiders to mate successfully. The eggs take about 9 weeks to hatch, watched over by the female during this period.

2.5 in. (6 cm)

Savannah

Terrestrial

Aggressive

Insectivorous

100–400 eggs

ZEBRA TARANTULA
Young zebra tarantulas assume adult coloration from about the age of 6 months. Male zebra tarantulas like this one are black in color, in spite of their name.

INDIAN BLACK AND WHITE TARANTULA

Poecilotheria regalis

The patterning on this species is even more attractive than in the case of the zebra tarantula. There is usually a big demand for this Asiatic tarantula, however, and the price of adult stock is invariably high. It will certainly be cheaper, and often more satisfactory, as in the case of other species, to start off with spiderlings if these are available. Your stock will be young and should have good breeding potential, whereas in the case of adult tarantulas, it is impossible to age them with certainty. Their quarters must include branches for climbing purposes, and some cork bark can be mounted at the back of the enclosure. They will build a retreat here. Their bite can be more serious than that of other tarantulas.

REPRODUCTION Spiderlings will grow fast on a diet of smaller crickets, moths, and other winged invertebrates. Spraying their quarters each day provides essential fluid for them. These spiders, and related tarantulas from this part of the world such as the Sri Lankan species (*P. fasciata*), are becoming established in collections.

2 in. (5 cm)

Tropical rainforest

Arboreal

Aggressive

Insectivorous

100–400 eggs

INDIAN BLACK AND WHITE TARANTULA
In spite of their name, yellow markings are also present on these spiders.

GIANT MILLIPEDES

family Sphaerotheriidae

A number of different giant millipedes occur in tropical parts of the world, but they all require similar care. Despite their name, their leg count rarely exceeds 240. They are protected by an unpleasant fluid that is produced by glands on the sides of the body. Handle them with care, and wear disposable gloves, particularly if you have any cuts on your hands. A much simpler method is to persuade the millipede to enter a clean, empty container such as a margarine tub, and then put on the lid.

There will be little need to disturb these millipedes once they are established in their quarters. Vermiculite, with leaf litter and tree bark on top, and some patches of moss, provides a suitable environment. An occasional light spray of water is recommended, and a water dish should be provided. Feeding these creatures is straightforward, as they will consume vegetable remains and sometimes even fruit, but take care that such foods do not become moldy in the vivarium.

11 in. (28 cm)

Tropical rainforest

Terrestrial

Agree in groups

Vegetarian

100 eggs

GIANT MILLIPEDES
These millipedes tend to hide away in their quarters.

REPRODUCTION Male giant millipedes have specially developed legs that are used in the mating process to pass the sperm into the female. Their genital openings are on the third segment of the body, near the fourth pair of legs. The male lies upside down to mate. The eggs are laid in a nest dug by the female, which may be lined with her excrement. This apparently helps to protect against fungal attack in damp surroundings. Hatching may take 3 or 4 weeks.

EMPEROR SCORPION

Pandinus imperator

6 in. (15 cm) 

Tropical rainforest 

Terrestrial 

Agree in groups 

Insectivorous 

3–90 live young 

Scorpions can be found in a wide range of habitats in the wild, often in deserts. This species is a denizen of the West African rainforest, however, and needs to be accommodated accordingly. Keepers of scorpions and similar invertebrates, including tarantulas, often use vermiculite to retain humidity in the vivarium. This inert substance, sold in garden centers and similar outlets, holds water well after being watered. In this case, a thin bark mix should be sprinkled on top of the vermiculite, and cover should be provided using cork bark or similar items.

Some scorpions are deadly, but generally, those with the largest pincers, such as the emperor scorpion, have a far less powerful sting in their tails. Even so, they should be handled carefully, as they can impart a sting of similar strength to that of a bee, which can be dangerous to those who are allergic to such stings.

It also makes sense to buy only from an experienced and knowledgeable dealer, so that you can be certain of the species which you are purchasing. Emperor scorpions can vary quite widely in coloration in any event, from shades of brown through to black. Their claws, or pedipalps, are covered with slight swellings, and they are also slightly hairy.

Feeding presents no difficulties, with invertebrates such as mealworms or wax-moth larvae being greedily snapped up. Avoid overfeeding, which can be fatal. The plates on the back of the scorpion's abdomen will start to lift if it becomes obese. Normally, scorpions need only be fed once every 2 weeks or so. A healthy individual is likely to rush out and grab its meal, which it then devours quickly.

EMPEROR SCORPION
Scorpions are one of the oldest life forms on the planet, having existed as a group for over 400 million years.

REPRODUCTION The breeding habits of scorpions are fascinating. It is not easy to separate the sexes reliably at first. Males may be slightly slimmer, with longer tails, but this is not an absolute means of distinction. Scorpions have a rather distant method of mating. The male drops a packet of sperm that is retrieved by the female when he pulls her to the spot, in what is sometimes described as the scorpion's dance. She will then give birth to live young, defending her offspring against any danger, and carrying them around on her back for the first week or so. If any fall off, she will stop and try to retrieve them. It can be possible to raise the young in groups, but other members of the colony may sometimes prey on them. It may be 6 years before they attain maturity, but they can live for 13 years or more.

TIPS

Don't use bug killers or other similar sprays close to a vivarium. The chemicals could kill the occupants very quickly.

A snake's eyes will turn milky when it is about to slough its skin. This is normal. A snake which sheds it skin whole is likely to be in good health, but if these so-called "spectacles" are not lost at this stage, they will ultimately need to be removed, so consult a specialist reptile veterinarian.

Take great care when using disinfectants, as some could be toxic. Stick to those brands recommended for use with reptiles, and follow the instructions for use carefully.

Incorporating an alarm system into the heating circuitry in the vivarium can alert you to any problem before it endangers the occupants.

Be sure to thaw out dead rodents and other foods of this type thoroughly, before offering them to reptiles or amphibians. Microwaves may not defrost them evenly, leaving ice crystals in the body. Special rapid thaw plates provide a better option.

Excessive lighting often causes turtles to develop algal growth on their shells. This should be wiped off regularly, as it can cause damage, growing under the edges of the scutes and causing them to lift up.

Always wash your hands immediately after attending to the needs of any creature.

If exercising your lizard on a leash, be sure to keep it away from other pets such as dogs. Both could be injured in any confrontation.

When using forceps to offer food, be sure these are blunt-ended rather than pointed, to minimize the risk of injury.

Do not use more than one type of dietary supplement at a time – otherwise, there is a real risk of overdosing the creature. Especially in the case of complete foods, additional supplements may not be recommended. If in doubt, consult your veterinarian.

GLOSSARY

Amphibian
A group of vertebrates whose reproductive biology is closely tied to water. Includes frogs, toads, newts and salamanders.

Aquaterrarium
A set-up which includes both an area of dry land and an expanse of water for swimming, as required by most amphibians and aquatic chelonians.

Arachnid
A group of invertebrates which includes spiders, scorpions, mites and ticks. Distinguishable from insects by having four pairs of legs.

Arboreal
Creature which spends part or most of its time off the ground, in trees for example.

Black light
A fluorescent tube valued in vivarium surroundings for its ultra-violet output, rather than for illuminating the occupants.

Bromeliads
A group of plants which are epiphytes, growing off the ground on trees. Used to decorate vivaria, with their central rosette of water being used as a breeding site for poison arrow frogs.

Carapace
The upper part of a chelonian's shell.

Chelonians
The reptilian group comprised of turtles, terrapins and tortoises.

CITES
The Convention on International Trade in Endangered Species, which regulates trade in wildlife.

Deworming
Giving treatment for roundworms and tapeworms.

Femoral pores
The area of scales running down the underside of the upper part of the hind leg, corresponding to the femur bone, especially significant for sexing lizards.

Fluffs
Young dead mice, with fur showing through their skin, used as reptile food.

Heaterstat
A combined heater and thermostat unit, used to heat water.

Heat pad
A slim-line heater, operated under thermostatic control, placed either under or on one of the outer sides of the vivarium.

Herbivorous
A creature which feeds primarily on foods of plant origin.

Hermaphrodite
A creature which has both male and female sex organs present in its body.

Herpetoculture
The keeping and breeding of reptiles and amphibians under controlled conditions.

Herptiles
A combined name for the reptile and amphibian groupings.

Hibernation
A period of inactivity usually triggered in response to a fall in temperature.

Hoppers
Immature form of locusts or crickets.

Insect
An invertebrate, whose body is comprised of three segments and six legs.

Insectivorous
A creature which feeds mainly on invertebrates.

Invertebrate
A creature without a backbone.

Larva
Immature stage in the lifecycle of some amphibians and invertebrates, hatching from an egg, and ultimately undergoing a change in form to become an adult.

Metamorphosis
The process whereby a larva changes into an adult.

Molt
Shedding of skin, often linked with growth in the case of invertebrates.

Mouth rot
Infected area in the mouth seen in some reptiles in poor condition.

Nymph
An immature but recognisable form of an adult invertebrate.

Palpal bulbs
Swellings on the front legs of male spiders.

Parthenogenetic
Invertebrate able to reproduce clones of itself without mating. A phenomenon associated with some stick insects.

Pinkies
Newly-born mice, used as food for reptiles and amphibians. Smallest size available.

Plastron
Underside of a chelonian's shell.

Quadruped
A creature with four limbs.

Semi-aquatic
A creature which spends part of its time on land, and also lives in the water.

Shedding
The term used to describe the molting of a snake's skin.

Snout
The nasal area.

Spawn
The eggs of amphibians

Spectacles
The old covering over the eyes, normally molted by snakes when they shed their skins.

Spiderlings
Immature spiders.

Supplements
Additives to the diet, usually comprised of vitamins and minerals. Used to compensate for likely deficiencies in the creature's diet.

Tadpoles
Larval form in an amphibian's life cycle.

Ultraviolet light
Component of sunlight which triggers synthesis of Vitamin D3 in the skin, and can also act as an appetite stimulant in many reptiles.

Venomous
Produces poison.

Vivarium
The enclosure in which herptiles or invertebrates are housed.

Zoonosis
A disease which can be spread from animals to people (and less commonly, vice-versa).

INDEX

Page numbers in *italics* refer to illustrations in the introductory section. **Bold** numbers refer to main entries in the Directory section, which include illustrations.

A

African bullfrog **121**
African clawed frog **103**
Agalychnis moreletii (Red-eyed tree frog) **111**
Agama **66**
Agama agama (Common agama) **66**
Alpine newt **132**
Amboina box turtle **98**
Ambystoma
  macrodactylum (Long-toed salamander) **127**
  maculatum (Spotted salamander) **129**
  mexicana (Axolotl) **130**
  opacum (Marbled salamander) **128**
  tigrinum (Tiger salamander) **127**
    mavortium (Barred tiger salamander) 127
Ameiva **48**
Ameiva ameiva (Ameiva) **48**
American bullfrog **120**
American gray tree frog **111**
American green toad **115**
American green tree frog **110**
American racer **76**
American toad **114**
amphibians 10–11, 100–37
  breeding 10–11, *10*, 39
  catching 101
  climbing methods 103
  conservation controls 40
  food 101
  handling 11
  health 34, 36
  skeleton *101*
  warning coloration 11
Anguis fragilis (Slowworm) 46, **49**
Anoles **62–3**
Anolis
  carolinensis (Green anole) **62**
  equestris (Knight anole) **63**
  sagrei (Brown anole) **63**
Aphonopelma
  chalcodes (Mexican blond tarantula **148**
  seemanni (Zebra tarantula) **149**
aquaterrarium, heating 16
aquatic environment
  semi-aquatic 26, *27*
  tropical 28, *29*
Arachnids 139
Asian tree frog **113**
Asian water dragon **50**
Asiatic horned toad **121**
Australian red-back tarantula 147
Avicularia avicularia (Pink-toed tarantula) **147**
Axolotl **130**
  vivarium environment for 28, *29*

B

Baird's rat snake 82
Ball python **72**
Barred tiger salamander 127
Bearded dragon **65**
Bell's hingeback tortoise **88**
Black rat snake **82–3**
Blotched kingsnake *30*
Blue-tongued skink, Eastern **54**
Boa constrictor (Common boa) 46, **71**
Boaedon fuliginosus see Lamphrophis fuliginosus
Boas 46, **68–72**, **71**
Bombina
  bombina (European fire-bellied toad) **119**
  orientalis (Oriental fire-bellied toad) **118**
  variegata (Yellow-bellied toad) **119**
Bosc's monitor **64**
Box turtles **98–9**
Brachypelma smithi (Mexican red-kneed tarantula) **146**
Brazilian rainbow boa 70
breeding 12–13, 38–9
Broken-striped newt 134
Brown anole **63**
Brown house snake **85**
Bufo
  americanus (American toad) **114**
  bufo, (Common European toad) **114**
  debilis (American green toad) **115**
  marinus (Giant toad) **117**
  punctatus (Red-spotted toad) **116**
  quercicus (Oak toad) **116**
  viridis (Green toad) **115**
  woodhousei (Woodhouse's toad) 114
Bull snake **76**
Bullfrogs **120–1**

C

Californian kingsnake 80
Carausius morosus (Indian stick insect) **141**
Caucasian crested newt 133
Centipedes 139
Ceratophrys ornata (Ornate horned frog) **105**
Chamaeleo
  jacksoni (Jackson's chameleon) **51**
  pardalis (Panther chameleon) **51**
Chameleons **51**
  handling *47*
Chelonians
  handling 47, *47*
  health 36, *37*
Chelonoidis
  carbonaria (Red-legged tortoise) **89**
  denticulata (Yellow-footed tortoise) 89
Chelus fimbriatus (Mata-mata) **97**
Chelydra serpentina (Common snapping turtle) **92**
Chicken snakes 82
Children's python **73**
Chilean pink tarantula 147
Chilean rose tarantula **147**
Chinese crocodile lizard **59**
Chondropython viridis (Green tree python) **73**
Chrysemys
  picta (Painted turtle) **93**
    belli 93
    dorsalis 93
    marginata 93
    picta 93
  scripta elegans (Red-eared turtle) **94–5**
Chuckwalla **61**
CITES (Convention of International Trade in Endangered Species) 40–1
Clawed frogs **102–3**
Cnemidophorus sexlineatus (Six-lined racerunner) **58**
Collared lizard **60**
Coluber constrictor (American racer) **76**
Common agama **66**
Common boa **71**
Common box turtle **99**
Common European toad **114**
Common garter snake **79**
Common kingsnake **80**
Common map turtle **93**
Common musk turtle **96**
Common snapping turtle **92**
Common wall lizard **55**
conservation controls 40–3
Corallus caninus (Emerald tree boa) **72**
Corn snake **84**
Couch's spadefoot toad **113**
Crevice creeper **112**
Crocodile lizard, Chinese **59**
Crocodile newt **137**
Crocodilions 43
Crotaphytus collaris (Collared lizard) **60**

Cryptobranchus alleganiensis (Hellbender) **122**
Cuora amboinensis (Amboina box turtle) **98**
Cynops pyrrhogaster (Japanese fire-bellied newt) **137**

D
dangerous animals 42–3
Danube crested newt 133
Dasypeltis scabra (Egg-eating snake) *75*
Day gecko, Giant **52**
dechlorinator 129
Dendrobates
  azureus (Blue poison arrow frog) *108*
  histrionicus 109
Dendrobatidae **108–9**
desert environment 18, *19*
Desert iguana **58**
Desmognathus ochrophaeus (Mountain dusky salamander) **126**
deworming 34
Diapheromera femorata (North American walking stick) **140**
dietary supplements 33, 153
Dipsosaurus dorsalis (Desert iguana) **58**
diseases 34–7
  fungal diseases 35
  mouth rot 37
  red leg 36
  salmonellosis 35
  tumors/lumps and bumps 36
Drymarchon
  corais (Indigo snake) **86**
    couperi 86
Durango mountain kingsnake 81
Dwarf clawed frog **102**
Dyscophus guineti (Southern tomato frog) **106**

E
Eastern blue-tongued skink **54**
Eastern kingsnake 80
Eastern mud turtle **96**
Egg-eating snake *75*
Elaphe
  guttata (Corn snake) **84**
    emoryi (Great Plains corn snake) 84
    rosacea (Rosy corn snake) 84
  obsoleta (Black rat snake) **82–3**
    bairdii (Baird's rat snake) 82
    lindheimeri (Texas rat snake) 82
    quadravittata (Yellow rat snake) 82
    rossalleni (Everglades rat snake) 82
    spiloides (Gray rat snake) 82
    williamsi (Gulf Hammock rat snake) 82
  schrencki (Russian rat snake) **84**
Emerald tree boa **72**
Emperor scorpion **151**
endangered species 40–1
Epicrates
  cenchria (Rainbow boa) **70**
    cenchria (Brazilian rainbow boa) 70
  striatus (Haitian boa) **70**
Eryx conicus (Rough-scaled sand boa) **68**
Eublepharis macularius (Leopard gecko) **52**
Eumeces fasciatus (Five-lined skink) **67**
Eurasian green tree frog **110**
European common frog **104**
European common newt **132**
European fire-bellied toad **119**
European toad **114**
Eurycantha calarata (Giant spiny stick insect) **144**
Eurycea
  longicauda (Long-tailed salamander) **125**
    guttolineata (Three-lined salamander) 125
Everglades rat snake 82
Extatosoma tiaratum (Giant prickly stick insect) **144**

F
feeding 30–3, 153
Fence lizard, Western **61**
Fire salamander **123**
Fire-bellied newt **137**
Fire-bellied toads **118–19**
Five-lined skink **67**
Florida kingsnake 80
Florida soft-shell turtle **97**
Four-toed salamander **126**
Frogs 38, 39, *100*, **102–13**, **120–1**
  distinction between toads and 100
  handling *100*
  health 36

G
Garter snake **79**
Geckos 8, 16, **52–3**
Gekko gecko (Tokay gecko) **53**
Geochelone pardalis (Leopard tortoise) **88**
Gerrhonotus multicarinatus (Southern alligator lizard) **49**
Giant day gecko **52**
Giant land snails, reproduction 13
Giant millipedes 150
Giant prickly stick insect **144**
Giant spiny stick insect **144**
Giant toad **117**
Golden mantella **107**
Goliath bird eater **148**
Gopher snake **76**
Grammostola
  cala (Chilean rose tarantula) **147**
  spatulata 147
Graptemys geographica (Common map turtle) **93**
Gray rat snake 82
Gray-banded kingsnake 81
Great Plains corn snake 84
Green anole **62**
Green iguana **56–7**
Green snake, Smooth **78**
Green toad **115**
Green tree python **73**
Gulf Hammock rat snake 82

H
Haitian boa **70**
health care 34–7
heaterstat 16
heating 16, *16*, *17*
Hellbender **122**
Hemidactylium scutatum (Four-toed salamander) **126**
Heosemys spinosa (Spiny turtle) **91**
heterodon nasicus (Western hog-nosed snake) **77**
Heteropteryx dilatata (Jungle nymph) **145**
Hingeback tortoise, Bell's **88**
Hog-nosed snake, Western **77**
Honduran milk snake 80
Horned frog **105**
Horned toad **121**
Horsfield's tortoise **91**
Hyla
  arborea (Eurasian green tree frog) **110**
  cinerea (American green tree frog) **110**
  versicolor (American gray tree frog) **111**
Hymenochirus boettgeri (Dwarf clawed frog) **102**

I
Iguana iguana (Green iguana) **56–7**
Iguanas 17, 31, **56–8**
incubator 39
Indian black-and-white tarantula **149**
Indian stick insect **141**
Indigo snake **86**
insects 138
  anatomy *138*
invertebrates 12–13, 138–51
  buying 13

J
Jackson's chameleon **51**
Japanese fire-bellied newt **137**
Jungle nymph **145**

K
Kaloula pulchra (Painted frog) **105**
Kingsnakes *30*, **80–1**, **87**
Kinixys belliana (Bell's hingeback tortoise) **88**
Kinosternon subrubum (Eastern mud turtle) **96**
Knight anole **63**

L
Lacerta viridis (Green lizard) **55**
Lamphrophis fuliginosus (Brown house snake) **85**
Lampropeltis
  alterna (Gray-banded kingsnake) 81
  calligister (Prairie kingsnake) **87**
    calligister 87
    rhombomaculata (Mole snake) 87
  getulus (Common kingsnake) **80**
    californiae (Californian kingsnake) 80
    florindana (Florida kingsnake) 80
    getulus (Eastern kingsnake) 80
  mexicana (Mexican kingsnake) **81**
    geeri (Durango mountain kingsnake) 81
    mexicana (San Luis Potosi kingsnake) 81
    thayeri (Thayer's kingsnake) 81
  triangulum (Milk snake) *41*, **80**
    campbelli (Pueblan milk snake) 80
    hondurensis (Honduran milk snake) 80
Latrodecotus mactans 147
Leopard frog **104**
Leopard gecko **52**
Leopard tortoise **88**
Liasis childreni (Children's python) **73**
Lichanura
  trivirgata (Rosy boa) **69**
    gracia 69
    roseofusca 69
lighting *16*, 17, *17*, 153
Litoria caerulea (White's tree frog) **112**
Lizards *8*, 46, **48–67**
  breeding 8
  feeding 31, 33
  handling 46, *46*
  health 36, 37
  lighting for 17, *17*
  vivarium for 15
Long-tailed salamander **125**
Long-toed salamander **127**
Lucanus cervus (Stag beetle) **140**

M
Madagascan pink-winged stick insect **145**
Mantella aurantiaca (Golden mantella) **107**
Mantis religiosa (Praying mantis) *12*, *138*, **142–3**
Map turtle **93**
Marbled frog **106**
Marbled newt **135**
Marbled salamander **128**
Mata-mata **97**
Mediterranean spur-thighed tortoise **90**
Megophrys nasuta (Asiatic horned toad) **121**
metamorphosis, triggering 130
Mexican blond tarantula **148**
Mexican kingsnake **81**
Mexican red-kneed tarantula **146**
Milk snake *41*, **80**
Millipedes 139, **150**
minerals 17, 30, 33
Mites 34, 139
Mole snake 87
Monitors **64**
Mountain dusky salamander **126**
Mud turtle, Eastern **96**
Mudpuppy **122**
Musk turtle **96**

N
Necturus maculosus (Mudpuppy) **122**
neoteny 10
Newts 11, 38, 39, *100*, **132–7**
  distinction between salamanders and 100
Nile monitor **64**
North American walking stick **140**
Northern crested newt **133**
Notophthalmus
  viridescens (Red-spotted newt) **134**
    dorsalis (Broken-striped newt) 134

O
Oak snake 82
Oak toad **116**
Opheodrys vernalis (Smooth green snake) **78**
Oriental fire-bellied toad **118**
Ornate horned frog **105**

P
Painted frog **105**
Painted turtles **93**
Pandinus imperator (Emperor scorpion) **151**
Panther chameleon **51**
parasites 34
parthenogenesis 12
Phelsuma madagascarensis (Giant day gecko) **52**
Phyllobates terribilis 108
Physignathus cocincinus (Asian water dragon) **50**
Pink-toed tarantula **147**
Pink-winged stick insect **145**
Pipa pipa (Surinam toad) **102**
Pituophis
  melanoleucus (Gopher snake) **76**
    sayi (Bull snake) **76**
Platysaurus guttatus (Red-tailed flat rock lizard) **67**
Plethodon 126
  cinereus (Red-backed salamander) **124**
  glutinosus (Slimy salamander) **124**
Podarcis muralis (Common wall lizard) **55**
Poecilotheria regalis (Indian black and white tarantula) **149**
Pogona barbatus (Bearded dragon) **65**
Poison arrow frogs *42*, **108–9**
  handling 11
  vivarium environment for 24
poisons, produced by animals 11, 42–3, 108
Polypedates leucomystax (Asian tree frog) **113**
Prairie kingsnake **87**
Praying mantis *12*, *138*, **142–3**
Pseudotriton ruber (Red salamander) **125**
Pueblan milk snake 80
Python
  regius (Ball python) **72**
  reticulatus (Reticulated python) **74–5**
Pythons **72–5**
Pyxicephalus adspersus (African bullfrog) **121**

R
Racerunner, Six-lined **58**
Rainbow boa **70**
Rana
  catesbiana (American bullfrog) **120**
  pipens (Leopard frog) **104**
  temporaria (European common frog) **104**
Rat snakes **82–3**, **84**
Red salamander **125**
Red-backed salamander **124**
Red-banded crevice creeper **112**
Red-eared turtle **94–5**
Red-eyed tree frog **111**
Red-kneed tarantula **146**
Red-legged tortoise **89**
Red-spotted newt **134**
Red-spotted toad **116**
Red-tailed flat rock lizard **67**
reptiles 8–9, **46–99**
  breeding 8, 39
  buying 9
  conservation controls 40
  dietary supplements 33

feeding 31
habitats 8
health 34, 35, 37
lighting for 17, *17*
transporting 9
vivarium environments for 18, *19*, 20, *21*
Reticulated python **74**
Rhacophorus leucomystax see Polypedates leucomystax
Ribbon snake, Western **78**
Rock lizard, Red-tailed flat **67**
Rosy boa **69**
Rosy corn snake 84
Rough-scaled sand boa **68**
Rough-skinned newt **136**
Russian rat snake **84**

S
Salamanders 39, *100*, **122–31**
distinction between newts and 100
vivarium environment for 22, *23*
Salamandra salamandra (Fire salamander) **123**
San Luis Potosi kingsnake 81
Sand boa, Rough-scaled **68**
Sauromalus obesus (Chuckwalla) **61**
savannah environment 20, *21*
Scaphiophryne marmorata (Marbled frog) **106**
Scaphiopus couchii (Couch's spadefoot toad) **113**
Sceloporus occidentalis (Western fence lizard) **61**
Scorpions 13, 139, *139*, **151**
semi-aquatic environment 26, *27*
Shinisaurus crocodilurus (Chinese crocodile lizard) **59**
Sipyloidea sipylus (Madagascan pink-winged stick insect) **145**
Six-lined racerunner **58**
Skinks **54**, **67**
Slimy salamander **124**
Slowworm 46, **49**
Smooth green snake **78**
Snails, reproduction 13
snakes 38, 46, **68–87**
breeding 8
determining age 87
feeding 33
handling 47, *47*
health 37
life expectancy 87
sloughing skin 152, *152*
transporting 9, *9*
venomous 42–3
vivarium for 15
Snapping turtle **92**
Soft-shell turtle, Florida **97**
Southern alligator lizard **49**
Southern tomato frog **106**
Spadefoot toad, Couch's **113**
Spaerotheriidae 150
Spiders 39, *43*, 139, *139*
Spiny stick insect **144**
Spiny turtle **91**
Spotted salamander **129**
Stag beetle **140**
Sternotherus odoratus (Common musk turtle) **96**
Stick insects 12, 13, *13*, 31, 38, **140–1**, **144–5**
handling 143
Surinam toad **102**

T
Tarantulas 13, 38, 139, **146–9**
Taricha granulosa (Rough-skinned newt) **136**
temperate woodland environment 22, *23*
Terrapene carolina (Common box turtle) **99**
Terrapins 37, 47
Testudo
graeca (Mediterranean spur-thighed tortoise) **90**
horsfieldi (Horsfield's tortoise) **91**
Texas rat snake 82
Thamnophis
proximus (Western ribbon snake) **78**
sirtalis (Common garter snake) **79**
Thayer's kingsnake 81
Theraphosa leblondi (Goliath bird eater) **148**
thermometers 16, *16*
thermostats 16, *16*
Three-lined salamander 125
Ticks 34, 139
Tiger salamander **127**
Tiliqua scincoides (Eastern blue-tongued skink) **54**
Toads 38, **102**, **113–19**
distinction between frogs and 100
taming 117
vivarium environment for 22, *23*
Tokay gecko **53**
Tomato frog **106**
Tomato toad *42*
Tortoises 31, *43*, **88–91**
handling 47
health 34, 37
mating *39*
taming 99
Tree frogs **110–11**, **112–13**
climbing methods *10*, 103
Tree python, Green **73**
Trionyx ferox (Florida soft-shell turtle) **97**
Triturus
alpestris (Alpine newt) **132**
apuanus 132
carniflex 133
cristatus (Northern crested newt) **133**
dobrogicus (Danube crested newt) 133
karelini (Caucasian crested newt) 133
marmoratus (Marbled newt) **135**
vulgaris (European common newt) **132**
tropical aquatic environment 28, *29*
tropical woodland environment 24, *25*
Turtles 8, 33, 39, **91–9**, 153
vivarium for 14–15, 26, *27*
Tylotriton verrucosus (Crocodile newt) **137**

U
ultraviolet light 17

V
Varanus
exanthematicus (Bosc's monitor) **64**
nilaticus (Nile monitor) **64**
vitamins 17, 30, 33
vivarium 152
environments 18–29, *18–29*
heating 16, *16*, *17*, 153
lighting *16*, 17, *17*
location 15
types 14–15, *14*, *15*

W
Walking stick **140**
Wall lizard **55**
water, heating 16
Water dragon, Asian **50**
Western fence lizard **61**
Western hog-nosed snake **77**
Western ribbon snake **78**
White's tree frog **112**
Woodhouse's toad 114
woodland environment
temperate 22, *23*
tropical 24, *25*

X
Xenopus laevis (African clawed frog) **103**

Y
Yellow rat snake 82
Yellow-bellied toad **119**
Yellow-footed tortoise 89

Z
Zebra tarantula **149**

PICTURE CREDITS

Key: *a* above, *b* below, *r* right, *l* left, *m* middle, *c* center

Ace Photo Agency 55*b*,

David Alderton 8*l*, 9*b*, 31*bl*, 33*cl*, 34ar, 34*br*, 35*b*, 36*ar*, 36*c*, 38*l*, 38*cl*, 38*mc*, 39*al*, 39cl, 46*al*, 88*a*, 88b, 91*a*, 91*b*

A.P.B. Photographic 11*ar*, 114*a*, 132*a*

Ardea London Ltd 13*a* *(Pascal Goetgheluch)*, 132 *(Hans D. Dosserbach)*

Dr Alan Beaumont 119*ar*, 130, 135, 144*br*

Daybreak Imagery 127*br* *(Todd Fisher)*

The Imagebank 30*ar*, 51*ar*, 53*bl* **(James Carmichael)**, 69*br*, 74, 106*a*, 137*bl* **(James Carmichael)**

Brian Kenny 33*ar*, 48*b*, 56*br*, 63*b*, 65*b*, 106*b*, 111*b*, 118, 123*ar*, 127*ar*

Chris Mattison 16*l*, 17, 18*ac*, 20*br*, 24*l*, 24*mc*, 28, 32*c*, 37*bl*, 39*br*, 46*bl*, 47*ar*, 47*br*, 58*a*, 58*b*, 59*b*, 63*ar*, 67*a*, 67*b*, 68*a*, 70*br*, 73*bl*, 77*br*, 78*br*, 81*a*, 84*br*, 85*al*, 87*br*, 97*b*, 100*ar*, 100*cr*, 100*br*, 102*bl*, 102*ar*, 105*b*, 110*ar*, 113*a*, 115*br*, 116*bl*, 116*ar*, 119*bl*, 120*bl*, 131*br*, 137*ar*, 139*ar*, 139*br*, 147*a*, 147*b*, 149*br*, 151, 152*cl*, 152*br*

Darren C. Maybury Front cover: *cl*, *al*, Back cover:a, 32*br*, 42*l*, 107*b*, 138*b*, 142

Oxford Scientific Films 75*r*, 122*a*, 122b, 136, 148*a*, 149*l*,

Papilio Photography Front cover: cr, 2, 10*l*, 37*ar*, 42*ac*, 44*ar*, 46*ar*, 49*ar*, 50*br*, 51*bl*, 52*bl*, 64*b*, 66*ar*, 90, 104*a*, 112*br*, 115*a*, 133*br*, 141

Pictor International 1, 5*a*, 5*b*, 6, 8, 9*a*, 12*bl*, 30*b*, 41*al*, 44*bl*, 47*al*, 101*l*, 150*l*

Photo Nats Inc Back cover: b, 35*ar* **(Stephen Maka)**, 36*bl* **(Tim Daniel)**, 41*br*, 42*br*, 61*bl*, 70*ar*, 79, 86*l*, 94, 100*cl*, 104*b*, 108, 110*b*, 111*a*, 121*bl*

Dennis Sheridan Wildlife Photography 12*al*, 13*br*, 34*l*, 38*bl*, 98, 101*r*, 105*a*, 107*a*, 112*a*, 140*b*, 144*a*, 145*a*, 145*b*, 152*ar*

Simpson and Co Nature Stock 10*ar*, 11*bl*, 40*r*, 43*cl*, 49*b*, 71*b*, 76*ar*, 76*br*, 78*ar*, 80*bl*, 82, 84*ar*, 93*a*, 93*b*, 96*ar*, 96*b*, 97*ar*, 99*bl*, 103*ar*, 109*al*, 113*br*, 117*b*, 123*bl*, 124*b*, 125*a*, 125*b*, 126*a*, 126*b*, 128, 129, 134*ar*, 148*b*,

Unicorn Stock Photos Front cover: ar, bl, 31*ar*, 40*l*, 43*ar*, 47*bl*, 52*bl*, 55*a*, 57*a*, 64*ar*, 52*ar* **(Doug Adams)**, 54*bl*, 60*br*, 61*ar*, 62, 72*ar*, 72*br*, 73*al*, 80*ar*, 92, 114*r*, 124*a*, 134*b* **(Richard Baker)**, 140*ar*, 146*a*, 150*r* **(Richard Baker)**

Windrush 121*ar*